Present from Andy
R. Cannon.

BIOGEOCLIMATIC UNITS OF CENTRAL AND SOUTHERN VANCOUVER ISLAND

by

AF864362

K. KLINKA, F.C. NUSZDORFER,

Ministry of Forests, Vancouver Forest Region,
355 Burrard Street, Vancouver, B.C. V6C 2H1

and

L. SKODA

Canadian Cartographics Limited, 508 Clarke Road,
Coquitlam, B.C. V3J 3X2

Province of British Columbia
Ministry of Forests

PROPERTY OF
B.C.I.T.

Canadian Cataloguing in Publication Data
Klinka, K., 1937-
Biogeoclimatic units of central and southern Vancouver Island.

Bibliography: p.
ISBN 0-7719-8213-5

1. Bioclimatology - British Columbia - Vancouver Island. 2. Ecological surveys - British Columbia - Vancouver Island. 3. Forest ecology - British Columbia - Vancouver Island. I. Nuszdorfer, F.C., 1952- . II. Skoda, L. III. British Columbia. Ministry of Forests. IV. Title.

QH543.K4 581.5'264 C79-092160-X

© 1979 Copyright
Province of British Columbia
Ministry of Forests, Victoria, B.C. V8W 3E7

ABSTRACT

Ecosystematic classification at the biogeoclimatic level has been refined further in that a new category, the biogeoclimatic variant, is described. Also, the ecological significance of biogeoclimatic units is substantiated by statistical treatment of climatic data. It is repeatedly shown by the analysis that the zonal ecosystems are reliable predictors of regional climate. A comparison of the annual actual evapotranspiration of biogeoclimatic units, suggested as an index of plant activity, is carried out. The most characteristic features of each biogeoclimatic zone, subzone, and variant are described with the aid of photographs. The distribution of these units on central and southern Vancouver Island is shown on a map produced at the scale of 1:500 000.

TABLE OF CONTENTS

TABLE OF CONTENTS (Continued)

ACKNOWLEDGEMENTS

We wish to express our thanks to the personnel of the Vancouver Forest Districts and the Regional staff for their support and assistance in collecting data, mapping, and preparing this report. Special thanks are due to Dr. V.J. Krajina of the Department of Botany, University of British Columbia for his help in selecting characteristic species, and to Mr. A.R. Fraser of the Forest Service Research Branch for his assistance in evaluating the statistical analysis.

Acknowledgements for assistance and cooperation are especially due to the personnel of the Forest Service Research Branch involved in ecological programs, in particular to Dr. T.E. Baker, Dr. R. Annas, and Dr. J. Pojar for useful discussions and valuable comments on the classification and manuscript; to Ms. S. Phelps of the Faculty of Forestry, University of British Columbia, for data processing; to Ms. L. Martin of the Vancouver Forest Region for the preparation of data; to Ms. C. Henry for the draughting of figures; to the typists who prepared the manuscript; and to Ms. A. Hope and Ms. C. Lowe for the final typing.

INTRODUCTION

The general acceptance of an ecosystem approach to forest management as a method of resource planning and a means of intensifying silvicultural practices necessitates the provision of reference material for forest practitioners. The explanatory information in this report and the accompanying biogeoclimatic map of central and southern Vancouver Island provide such material.

This biogeoclimatic map is a forerunner of a provincial series at a scale of 1:500 000. It is a corollary to those of Krajina (1973) and MacMillan Bloedel (1974). On it, the distribution of biogeoclimatic zones, subzones, and variants and their relationship to other landscape features are shown. These units represent groupings (classes, taxa) of ecosystems in relation to regional climate. They are being used in an ecological stratification program of the province by the Research Branch of the B.C. Ministry of Forests. Biogeoclimatic units form a framework in guides for tree species selection and prescribed burning, serve as an index to ecosystem identification and interpretation, provide comparative benchmarks for further research, and facilitate communication.

This report briefly reviews principles of classification at the biogeoclimatic level. The classification of biogeoclimatic units in the Vancouver Forest Region follows. To make the reader familiar with the most characteristic features of the units, they are described briefly and illustrated with photographs. Differentiating characteristics of climatic data are identified using a statistical technique; and a brief explanation of the methods employed in mapping is included. Finally, the cartographic design, used by Canadian Cartographics Ltd., is described.

PRINCIPLES AND METHODS OF BIOGEOCLIMATIC CLASSIFICATION OF ECOSYSTEMS

Although ecosystematic (synecological) classification is increasingly used in forest management, it remains insufficiently understood. It is commonly thought to be a classification of vegetation only. The few published explanations of the system are those of Krajina (1959, 1965, 1969, 1972), Mueller-Dombois and Ellenberg (1974), and Beil *et al.* (1976). In this section the classification is reviewed and explained.

The Conceptual Basis

Both forestry and agriculture are based on plant production, but the former has a much longer period between harvests. Understanding and reliably predicting plant growth are, therefore, prerequisites to successful forest management. It is of particular importance to understand that forests are more than timber stands; they are interacting complexes of living organisms and the environment immediately affecting them. These complexes are termed ecosystems or biogeocoenoses.

In this report an ecosystem (Tansley 1935) and a biogeocoenose (Sukachev and Dylis 1964a, b) are considered synonymous. This is done to simplify terminology, acknowledging that the former has a more general definition than the latter. According to our translation of the definition given for a forest biogeocoenose by Sukachev and Dylis (1964a), it is 'that part of the forest uniform over a certain area in the composition, structure, and properties of its components, and in the interrelationships among them; that is, uniform in the plants, animals, and microorganisms inhabiting it, in the parent material, in its hydrological, microclimatic (atmospheric), and soil environments and the interactions among them; and in the kind of matter and energy exchange between these components and other natural phenomena in nature.' This agrees reasonably well with that rendered by Maclennan in his translation (Sukachev and Dylis 1964b).

Plants and soil are two important and discernible components of ecosystems. They integrate and reflect factors such as soil parent material, organisms, physiography, time, and climate. Thus, on the basis of plants and soil, uniformity is ascertained. Indeed, it would be ideal to quantify all the components of an ecosystem, but this is prohibitive in terms of time and sometimes impossible due to limitations in technology. Plants are considered uniform relative to their variation in the landscape. In this context plant cover is subdivided wherever there are recognizable changes in floristic composition, structure, and vigor. Soils are considered uniform if the kinds and arrangements of horizons are the same. This means the morphological characteristics of horizon color, texture, structure, consistence, thickness, reaction, and composition must fall within a narrow range. An area which is uniform in terms of plants and soil is the smallest thing dealt with in ecosystematic classification. Lateral boundaries between biogeocoenoses are either distinct or gradational, depending on the abruptness of the differences in the structure and properties of the components.

To cope with the large numbers of ecosystems, it is necessary to organize our knowledge about them so it can be utilized and communicated efficiently. A classification system does this while also providing a context in which to gather more information. It gives, so to speak, the table of contents and the chapter headings under which the science to be studied will be methodically divided, and indicates where the acquired knowledge is to be arranged under each of these headings (Duhem 1974). Given sufficient information about the ecosystems, it becomes possible to infer things about others which are similar but have not been studied in detail. This power of prediction is one of the most important aspects of ecosystematic classification.

According to Lavkulich (1972), there are both natural taxonomic and interpretive classifications. The first is objective while the second is subjective or technical. Ecosystematic classification is a natural taxonomic method, the objective of which is to show relationships between the most common and important properties of ecosystems (Cline 1949). First, classes are formed by grouping ecosystems which have similar selected properties; then they are named, defined, and organized. After establishing a solid framework with this scientific tool, the next step is to develop practical applications for the intensive management of forest resources. When ecosystems are grouped to serve such a purpose, the classification is an interpretive one. Ecosystematic classification tends to be relatively complex, containing numerous groups. The differences between some of these are small relative to the scale of management treatments, so for interpretive classification they are gathered into larger groups. It is to these larger groups that management prescriptions are applied. So, the techniques described in this report involve both ecosystematic and interpretive classifications.

Principles of Biogeoclimatic Classification of Ecosystems

Regional climate plays a major role in dictating the nature of ecosystems, its sphere of influence often covering large geographical areas. Thus, it is useful to create generalized groups with similar climate. This is done at the biogeoclimatic level of ecosystematic classification. Classification at this level would be difficult, particularly where climatic data are sparse or lacking entirely, if the units recognized were not linked functionally to ecosystems. This link is provided by the biogeocoenotic level of classification. Therefore, ecosystematic classification has biogeoclimatic and biogeocoenotic levels.

In relating climate to ecosystems, zonal ecosystems play a major role. In them the influence of climate is expressed most strongly. The predominant input of water is as precipitation. It becomes water which contains nutrients and other materials, stored in and moving through the soil. In zonal ecosystems the soil has an intermediate capacity to hold water, a condition often occurring when soil texture lies in the loam range and the materials are moderately deep. As a result, water and nutrients are available for plant growth for a longer period of time than would be the case in soils where there is a low water-holding capacity. In contrast, they are available for a shorter period of time than is the case in ecosystems with a higher water-holding capacity or where there are signficant additional inputs of moisture, nutrients, and materials as seepage or groundwater. Water, nutrients, and other materials leave zonal ecosystems as seepage and groundwater while only water leaves as evapotranspiration. For the intermediate soil described above, the relationship between inputs to, storage in, and losses from ecosystems that leads to zonal conditions occurs most frequently when they are located on the upper topographic positions of flat watershed divides or on the upper to middle parts of gentle slopes.

Plant growth in zonal ecosystems reflects the intermediate or mesic moisture and intermediate or mesotrophic nutrient regimes. Other ecosystems are influenced more strongly by local physiography (*e.g.* position on slope, slope gradient, and aspect) and the physical and chemical properties of parent materials. Therefore, they do not reflect the regional climate so clearly. Specifically, they may be drier or wetter and richer or poorer than zonal ecosystems.

The composition, structure, and vigor of plants in zonal ecosystems changes until a climatic climax state is attained, reflecting the development potential of the regional climate. In contrast, ecosystems that are not zonal will reach edaphic (related to physiography and soil moisture and nutrient regimes) climaxes. Theoretically, under similar successional and pedogenic processes, these latter climaxes will converge into the climatic climax. Usually, this does not occur due to counterbalancing events such as fire, flood, strong wind, disease, insect attack, and human disturbance.

An ecosystem is designated as climax if the species of plants are perpetuating on the site. Mortality and regeneration are restricted to individuals or small groups. As a result, climatic climax tree species may be present in herb, shrub, and all tree layers. Similarly, shrubs, herbs, mosses, liverworts, and lichens may be present in various stages of maturity. In the climax state the plants of an ecosystem have certain age, structural, and compositional characteristics, resulting in uninterrupted biogeochemical cycling and energy exchange and transfer.

In zonal ecosystems only tree species which tolerate shade and intermediate moisture and nutrient regimes regenerate continually under a forest canopy and, thus, perpetuate successfully. The same species may require shade in drier climates or be intolerant of it in more humid climates on the same kind of ecosystem. Thus, a tree species' tolerance to shade determines whether or not it will be present in successional or climax stages in a given ecosystem. The distribution of zonal ecosystems in the landscape can be used to help delineate areas with similar regional climates. For classification, it is desirable to use climatic climax zonal ecosystems because on these there has been sufficient time for the climate to exert its influence on the vegetation.

Methods of Classification

Krajina (1965, 1969, 1972) proposed four categories at the biogeoclimatic level: formation, region, zone, and subzone. Formations and regions unify zones by similarities in general characteristics of macroclimate, predominant soil processes, and physiognomic aspects of the upper vegetation strata of zonal ecosystems. Formations and regions are not affected by synecological studies being carried out within British Columbia. Zones and subzones have been found useful in both research and application; efforts are underway to complete classification at this level for major parts of the province.

Biogeoclimatic Units

Biogeoclimatic zones in British Columbia were derived by Krajina on the basis of regional climate and broad characteristics of zonal ecosystems and complemented by descriptions of prevailing pedogenic processes as well as characteristic plants. Zones were divided into subzones, which were recognized on the basis of more specific climatic features and subsets of the generalized groups of zonal ecosystems. These were also complimented by more detailed descriptions of pedogenic processes and lists of characteristic species.

The classification system was designed to be flexible so new information could be accommodated. For example, for the area covered in this report it became evident that the previously designated 'variation of a subzone' could be more adequately dealt with as a new category, the biogeoclimatic variant. Each biogeoclimatic subzone may be divided into several variants on the basis of further differentiation of its climate. For instance, obvious differences in climate are evident between low and high elevations of the Wetter Maritime Coastal Western Hemlock Subzone. This significantly affects the productivity of ecosystems. In general, biogeoclimatic variants are recognized when an area is drier, wetter, warmer, or colder than what is considered near average for the subzone.

Since the variations of subzone climate control development of numerous ecosystems over large continuous areas, they are still considered to be regional in extent. However, the change in zonal ecosystems between variants within a subzone is less obvious than the change in zonal ecosystems between subzones within a zone. In the plant community these are manifested as changes in its structure or in the proportion and vigor of certain species. Differences in soils appear to be confined to pedogenic processes and are not markedly expressed in morphological features. The categories described above, along with differentiating and accessory characteristics, are listed in Table 1.

In addition to the variant, the biogeoclimatic phase accommodates variations of regional climate resulting from local relief. Particularly in warmer and drier regional climates, the need for this becomes clear when zonal ecosystems on northern and southern exposures are compared. It can also be applied in mountainous terrain where cold-air drainage affects ecosyst development. Since the phase is related to a local climate, it is not employed in biogeoclimati classification. It is, nevertheless, useful in accounting for ecosystems that are atypical for a regional climate and yet are not a result of edaphic conditions.

Table 1. Differentiating and Accessory Characteristics at the Biogeoclimatic Level of Ecosystematic Classification

Biogeoclimatic Category	Differentiating Characteristics	Accessory Characteristics
Zone	ranges of climatic variables for the zone; zonal ecosystems classified at the level of plant order *	characteristic combination of species for the zone; pattern of plant orders occurring throughout the zone
Subzone	ranges of climatic variables for the subzone; zonal ecosystems classified at the level of biogeocoenotic association *	characteristic combination of species for the subzone; pattern of biogeocoenotic associations occurring throughout the subzone
Variant	ranges of climatic variables for the variant; zonal ecosystems classified at the level of biogeocoenotic type *	characteristic combination of species for the variant; pattern of biogeocoenotic types occurring throughout the variant

* these terms will be discussed in the following section.

In this report names of zones follow those of Krajina. They are usually based on the climatic climax tree species most commonly growing on zonal ecosystems. Subzones are named by adding descriptive terms to zone names. Variants are named by adding physiographic modifiers to the subzone name (after Holland 1976). On the map the biogeoclimatic units are coded by a simple alphabetic-numeric system. For example, the West Vancouver Island Drier Maritime Coastal Western Hemlock Variant is designated CWHa1, where the subzone modifier is 'Dry Maritime', the symbol is 'a', the variant modifier is 'West Vancouver Island', and the symbol is '1'.

Throughout the text, certain abbreviations in relation to zone names are used: Alpine Tundra = AT, Mountain Hemlock = MH, Coastal Western Hemlock = CWH, and Coastal Douglas-fir = CDF. The terms 'biogeoclimatic zone', 'biogeoclimatic subzone', and 'biogeoclimatic variant' are simplified to 'zone', 'subzone', and 'variant', respectively.

Biogeocoenotic Units

The biogeoclimatic level stratifies climate, while the biogeocoenotic level deals with vegetation and soil. In biogeocoenotic classification the association is the central unit. It is, essentially, identical to the plant association described by Krajina (1960). Associations are abstracted through synthesis of relevés from several to many sample plots. The differentiating characteristics for associations include the characteristic combination of species and values of soil moisture and nutrient regimes. Ecosystems that are associated are considered to undergo specific succession, culminating in climax plant communities. This is based on the premise that, under similar environmental conditions, succession of disturbed ecosystems will result in the development of similar climax ecosystems. It should be noted, however, that following disturbances the recurrence of identical ecosystems, particularly with respect to soils, cannot be expected. Associations which are related in floristic composition, successional trends, broad environment-vegetation relationships, and prevailing pedogenic processes are combined as plant orders.

Associations are also subdivided into biogeocoenotic types. These are identical to the type of biogeocoenosis of Sukachev (1944) and Sukachev and Dylis (1964a, b). Types have certain environmental, particularly edaphic, characteristics in common. Among these may be soil depth, particle size, horizon sequences, humus form, parent material, bedrock geology, and other physical and chemical properties affecting plant vigor. In terms of the vegetation of types within an association, there may be minor floristic differences evident in the layering of species or in their significance and vigor. Each type is characterized by the differentiating criteria of the association to which it belongs plus selected floristic, but mainly environmental features. To date only some taxonomic work has advanced to the stage of classifying types, so only a few taxa are defined.

The present vegetation cover may be very heterogeneous or lacking entirely in extreme cases because of various disturbances. Even considering only two key variables of tree layers - species composition and age - a great diversity of forest communities exists. To deal with this variability, types are subdivided into biogeocoenotic variants. These are the same as the syngenetic units of Braun-Blanquet (1932) and the derivative types of biogeocoenoses of Sukachev and Dylis (1964a, b). The category of the biogeocoenotic variant refers to various successional stages of plant communities and associated changes in biogeoclimatic cycling which may develop following the disturbance of a particular biogeocoenotic type. The differentiating characteristics for the type plus the age, density, and tree species composition of forest stands are included in characterizing biogeocoenotic variants. Very little taxonomic work at this level has been carried out so that numerous taxa are not described.

There is often no direct relationship between the moisture and nutrient regimes governing plant growth and the morphological criteria used for soil classification. For this reason it is often not possible to use single taxonomic units of soil for differentiating biogeocoenotic units, particularly when the level of soil series is encountered. For example, in an extreme case a mesic moisture and mesotrophic nutrient regime could develop in the Wetter Maritime CDF Subzone in soils classed to the Brunisolic and Podzolic orders.

A common and scientific nomenclature is employed for biogeocoenotic categories in this study. Names of orders and associations consist of the tree species preceded by selected characteristic species from moss-liverwort-lichen, herb, and shrub layers. Names of types include the association name and the name of the soil order, great group, or subgroup plus a description of the humus form. For example, the biogeocoenotic type integrating zonal ecosystems from the West Vancouver Drier Maritime CWH Variant is: Stokesiella - Rhytidiadelphus - Douglas-fir - Western Hemlock on Humo-Ferric Podzol with friable mor humus. The order of names of plants in biogeocoenotic units follows that of European phytosociologists such as Braun-Blanquet (1932), and emphasizes the importance of understory vegetation in a synecological classification by listing those species first.

To shorten the length of names of biogeocoenotic categories the following abbreviations are made: AA - *Abies amabilis*, AG - *Abies grandis*, AL - *Abies lasiocarpa*, AM - *Arbutus menziesii*, CN - *Chamaecyparis nootkatensis*, PS - *Picea sitchensis*, PC - *Pinus contorta*, PM - *Pseudotsuga menziesii*, QG - *Quercus garryana*, TP - *Thuja plicata*, TH - *Tsuga heterophylla*, and TM - *Tsuga mertensiana*. A complete list of scientific and common names is given in Appendix III.

Within a particular subzone or variant the pattern of biogeocoenotic associations or types, respectively, can be projected onto an edaphic grid (Figure 1). This technique is adapted from Krajina (1969). The generalized zonal ecosystems occupy the central or 4/C position on this grid. Other generalized ecosystems are placed on the grid according to their moisture and nutrient regimes. These values are determined both according to the range of ecosystems in an area under the influence of a regional climate and by comparison with the generalized zonal ecosystem.

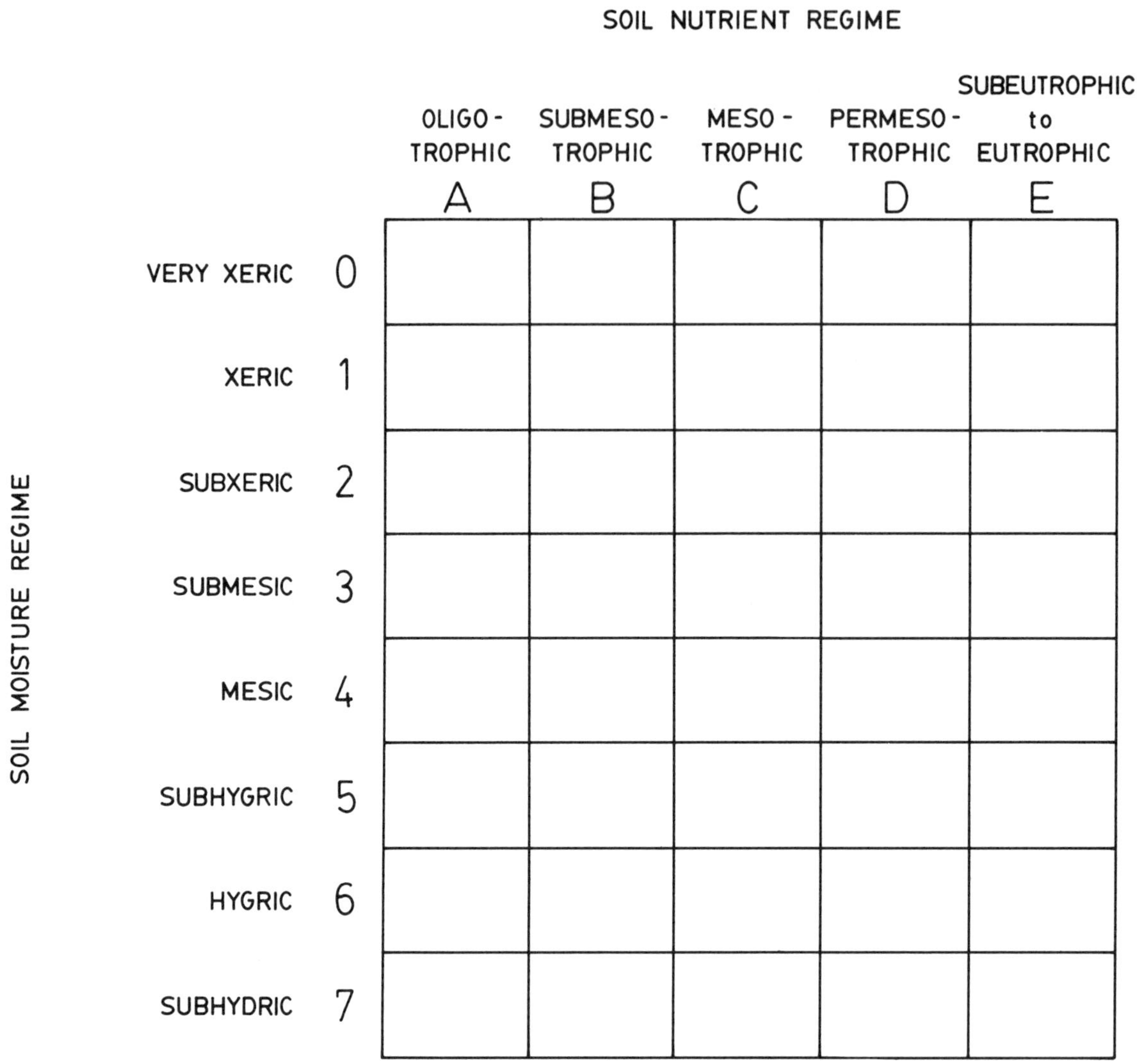

Figure 1. An edaphic grid of soil moisture and nutrient regimes.

Problems may arise in using such a simple tool for ranking ecosystems. For example, only a few zonal ecosystems may be found where there is a very moist regional climate and a subdued topography *i.e.* where there is not an excess of water present in the soil or organic matter. In such a case the upper part of the grid in Figure 1 would remain virtually empty and a different technique of ranking the generalized ecosystems could be required. This problem did not arise in the area dealt with in this report. Until considerably more research is done on the functional aspects of numerous ecosystems,there is no good alternative to the edaphic grid. Eventually, we hope that the relative 'functional' terms can be replaced by their real counterparts based on physical and chemical measurements of the soil.

Once this grid is established it has a number of applications: It facilitates communication; can be used for further classification; may be employed for cautious extrapolation of observations or experimental results to other ecosystems within the area of a regional climate; and information about generalized ecosystems from various cells may be used for specific purposes. An example of the last point is the projection of the data regarding shade tolerance of tree species within cells. If this is done for all the tree species, it is possible to determine successional trends for individual cells. With this framework it is possible by noting the soil moisture and nutrient regime of an ecosystem which is in an earlier successional state, to predict the eventual climatic climax tree species. For other applications the grid may be simplified by grouping cells together if they react similarly to certain treatments or practices.

Due to the rather limited data base, edaphic grids could not be decisively and consistently applied throughout the area of the Vancouver Forest Region. In this report the grid coordinates are employed in the illustration captions to describe the units in terms of soil moisture and nutrient regimes.

Characteristic Combinations of Species

Characteristic combinations of species are arranged alphabetically in the order of trees, shrubs, herbs, mosses - liverworts, and lichens. Plants found in zonal ecosystems were preferred in the selections since the regional climate affects them most strongly. Those plants tolerating wider ranges of soil moisture and nutrient regimes also occur in other ecosystems, but they are still included if their distribution is strongly related to climate. A plant species may be in more than one characteristic combination of species. Finally, the characteristic combination of species for a variant includes its species plus the combination for the subzone and zone; and for a subzone the combination of species for the zone is included.

The characteristic combinations of species listed are chosen on the basis of results from data analysis for numerous sample plots taken throughout the biogeoclimatic categories. By comparing zones, subzones, and variants it is possible to identify species with specific climatic tolerances. These plants may also occur in other categories - particularly where variants are concerned - but usually a difference in vigor or/and abundance is obvious. It must be borne in mind that in areas where one category borders another there may be admixtures of corresponding characteristic combinations of species. This is particularly evident where boundaries between zones, subzones, or variants are diffuse and cover considerable ground. In terms of forest management, this is not serious if a compromise is struck between the guides for treatment of the respective categories. Usually the recommendations will be similar. However, instances will occur when the forest manager must rely on knowledge of the area.

Discriminant Analysis of Climatic Variables

With discriminant analysis it is possible to mathematically select the climatic parameters that are acting on zonal ecosystems in such a way that the ecosystems develop unique characteristics within a biogeoclimatic category. In this statistical technique, the number of biogeoclimatic categories recognized on the basis of distribution of zonal ecosystems is also the number of groups used in the analysis. Each category is represented by an unique linear function. The functions are generated from the climatic parameters that best serve to distinguish between the categories.

To carry out the analysis the computer program, UBC BMD07M (Halm 1970), was employed. It does a stepwise discriminant analysis. In it, the climatic parameters are entered into linear functions in a stepwise manner. Each time a parameter is added a statistical test is done to determine if the distinction between the categories has increased. If its inclusion is not beneficial it is dropped, otherwise it is retained. The usefulness of each parameter is tested every time a new one is added; those no longer making a significant contribution are dropped. Parameters are ranked by the program according to their importance in maintaining the category separation. This order of importance is displayed in the tables throughout the text of this report with the exception of Appendix II. In addition, canonical variables for each parameter chosen for the discriminant functions are calculated and plotted to give a two-dimensional illustration of the degree of separation of groups. The calculation of canonical variables is from the statistics for the climatic parameters and can be used only on a relative basis. For specific formula concerning any steps described, the reader should consult the documentation referred to above.

A total of 138 observations, each with 32 variables, were provided for the analysis by the Resource Analysis Branch, Ministry of Environment. The observations consist of original and derived climatic data (Appendix I). The original data were from records of climatic stations maintained by the Federal Department of Environment, Atmospheric Environment Service; the derived data were obtained from the former using a remote station climate parameter prediction model (Rose and Grant 1976). The derived data were predicted for points in close proximity to respective climatic stations at altitudes representative for a biogeoclimatic unit. The predicated data become less accurate as elevation and distance from the original station increase. Furthermore, the model is still in the developmental stages. The reader is cautioned to view the climatic data presented in the text, Appendix II and Appendix IV with this in mind.

Without the derived data, the analysis and climatic characterization of montane and subalpine biogeoclimatic units would not be possible. The variables selected are those used by Köppen and Trewartha (Trewartha 1957), Krajina (1959, 1965, 1969), and Major (1963).

The index of continentality was calculated as follows:

$$IC = 1.7\,(\text{Tmean (July)} - \text{Tmean (January)}) \div \text{sine (degrees latitude)} - 20.4$$

where IC = index of continentality

T = temperature.

Equilibrium evapotranspiration was calculated according to the method of Wilson and Rouse (1972). This value, multiplied by 1.26, gives an estimate of equilibrium potential of evapotranspiration, referred to in this report simply as potential evapotranspiration. There is evidence that this constant may be approximately 25% too high (McNaughton and Black 1973; Black, in press). However, sufficient information from various forest stands under different climatic regimes precludes the use of a different method of calculation at this time.

Actual evapotranspiration was calculated following Thornthwaite's (1948) procedure with 100 mm as the value for available water stored in the soil. The method involves balancing precipitation, soil storage, and potential evapotranspiration on a monthly basis. If precipitation plus the water remaining in the soil from the previous month is greater than or equal to potential evapotranspiration, then actual evapotranspiration is equal to potential evapotranspiration. If precipitation plus the water remaining in the soil from the previous month is less than potential evapotranspiration, then actual evapotranspiration is equal to precipitation plus the water remaining in the soil from the previous month. Water stored in the soil decreases by the amount that potential evapotranspiration exceeds precipitation. When water stored in the soil drops to 0 mm, the deficit is equal to the amount potential evapotranspiration exceeds precipitation. Recharge is equal to the amount precipitation exceeds potential evapotranspiration. The summation of monthly values gives the figures on an annual basis. Actual evapotranspiration is employed in discriminant analysis and also as an index of zonal plant activity under natural conditions as suggested by Major (1963). A summary of climatic parameters appears in Appendix II.

Throughout tables in the text and in Appendix II, the climate is coded according to the system of Köppen, as described in Trewartha (1968). The symbols are explained in this reference and summarized below:

C = Mild temperate rainy climates; coldest month mean temperature is between -3 and 18°C; warmest month mean temperature is over 10°C;

s = dry summer; wettest winter month has at least three times the precipitation of the driest summer month; less than 30 mm of precipitation occurs in the latter;

f = no distinct dry season; wettest winter month has less than three times the precipitation of the driest summer month; more than 30 mm of precipitation occurs in the latter;

b = cool summer; mean temperature of warmest month is under 22°C;

c = cool short summer; fewer than 4 months have mean temperature greater than 10°C;

D = cold snowy climates; coldest month mean temperature is below -3°C; warmest month mean temperature is over 10°C;

f = humid winters;

c = as defined for c above.

BIOGEOCLIMATIC UNITS ON THE NOOTKA-NANAIMO MAP SHEET

A synopsis of biogeoclimatic units recognized in the Vancouver Forest Region is given in Table 2. The units are essentially those of Krajina and his students, and include the most recent revision of biogeoclimatic zonation (Krajina 1975b, 1978). In addition, the synopsis presents several new subzones and variants, particularly those having maritime (oceanic) climate. The classification of the subcontinental (or so-called transitional) subzones and the Wetter Maritime Northern CWH Subzone is still under study.

The differentiating characteristics for the biogeoclimatic units that occur in the area of the Nootka-Nanaimo map are listed in Table 3. One of the accessory characteristics for these units is a characteristic combination of species (Table 4). Differentiating climatic characteristics for biogeoclimatic units are presented in tables throughout the text. All of these tables together represent the reference material for each biogeoclimatic unit discussed. In addition, a cross section showing a generalized sequence of zonal ecosystems for variants is illustrated on the map.

The entire set of climatic data from the CDF, CWH, and MH zones was subjected to discriminant analysis. The climatic data conforms closely to the three zones with 96.4% of the station locations or prediction points classified correctly. Fourteen variables were selected as contributing significantly to the differentiating functions. They are located in order of decreasing importance from top to bottom in Table 5. Climatic data from other stations can be classified to a particular zone using the linear functions in Table 6.

The plot of canonical variables shows the relationship between the CDF and CWH zones (Figure 2). During mapping a diffuse boundary was encountered, especially along the latitudinal gradient in the northern limits of the CDF Zone. The two zones were grouped by Krajina into the next higher category - the Pacific Coastal Mesothermal Forest Region. Looking again at Figure 1, it is apparent that the climatic observations from the MH Zone are distinctly separate from the others, indicating little affinity between the ecosystems of the Pacific Coastal Mesothermal and Subalpine Forest Regions.

From the curves of calculated actual evapotranspiration for the CDF, CWH, and MH zones (Figure 3) a number of interpretations are possible. First, in reponse to low temperatures,

Table 2. Synopsis of Biogeoclimatic Units in the Vancouver Forest Region.

Formation	Region	Zone	Subzone	Variant
Arctic or Alpine	Alpine	Alpine Tundra (AT)	Coastal AT (ATa)	
			Coastal-Interior AT (ATb)*	
Microthermal Coniferous Forest	Pacific Coastal Subalpine Forest	Mountain Hemlock (MH)	Maritime Forested MH (MHa)	West Vancouver Island Maritime Forested MH (MHa_1)
				East Vancouver Island Maritime Forested MH (MHa_2)
				Pacific Ranges Maritime Forested MH (MHa_3)
			Maritime Parkland MH (MHap)	
			Subcontinental Forested MH (MHb)*	(in study)
			Subcontinental Parkland MH (MHbp)*	
	Canadian Cordilleran Subalpine Forest	Engelmann Spruce - Subalpine Fir (ESSF)*	(in study)	
	Canadian Cordilleran Forest	Interior Douglas-fir (IDF)*	(in study)	
Mesothermal	Pacific Coastal Mesothermal Forest	Coastal Douglas-fir (CDF)	Drier Maritime CDF (CDFa)	Nanaimo & Georgia Drier Maritime CDF ($CDFa_1$)
			Wetter Maritime CDF (CDFb)	Nanaimo & Georgia Wetter Maritime CDF ($CDFb_1$)
				Fraser Wetter Maritime CDF ($CDFb_2$)*
			Wetter Subcontinental CDF (CDFc)*	(in study)

Coastal Western Hemlock (CWH)	Drier Maritime CWH (CWHa)	West Vancouver Island Drier Maritime CWH ($CWHa_1$)
		East Vancouver Island Drier Maritime CWH ($CWHa_2$)
		Pacific Ranges Drier Maritime CWH ($CWHa_3$)
	Drier Subcontinental CWH (CWHc)*	(in study)
	Wetter Maritime CWH (CWHb)	Estevan Submontane Wetter Maritime CWH ($CWHb_1$)
		West Vancouver Island Submontane Wetter Maritime CWH ($CWHb_2$)
		West Vancouver Island Montane Wetter Maritime CWH ($CWHb_3$)
		East Vancouver Island Submontane Wetter Maritime CWH ($CWHb_4$)
		East Vancouver Island Montane Wetter Maritime CWH ($CWHb_5$)
		Pacific Ranges Submontane Wetter Maritime CWH ($CWHb_6$)*
		Pacific Ranges Montane Wetter Maritime CWH ($CWHb_7$)
	Wetter Subcontinental CWH (CWHd)*	(in study)
	Wetter Maritime Northern CWH (CWHe)*	(in study)

* Ecosystems of this biogeoclimatic unit do not occur in the area covered by the Nootka-Nanaimo map sheet.

Table 3. Synopsis of Zonal Ecosystems Employed as Differentiating Characteristics for Biogeoclimatic Units - Nootka-Nanaimo Map Sheet

Biogeoclimatic Zone	**Plant Order**	Biogeoclimatic Subzone	Biogeocoenotic Association	Biogeoclimatic Variant	Biogeocoenotic Type
Alpine Tundra (AT)	Mountain Heather	Coastal AT	**Pink & Mertens' Mountain Heather**	not differentiated	
Mountain Hemlock (MH)	Black Blueberry - Mountain Hemlock	Maritime Forested MH	Alaskan & Black Blueberry - Amabilis fir - Mountain Hemlock	West Vancouver Island Maritime Forested MH	Oval-leaved & Alaskan Blueberry Amabilis Fir - Mountain Hemlock on Ferro-Humic or Humic Podzols with compacted mycelial mor humus
				East Vancouver Island Maritime Forested MH	Oval-leaved & Black Blueberry Amabilis Fir - Mountain Hemlock on Ferro-Humic Podzol with compacted mycelial mor humus
				Pacific Ranges Maritime Forested MH	Alaskan & Black Blueberry - Amabilis Fir - Mountain Hemlock on Ferro-Humic or Humic Podzols with compacted mycelial mor humus
		Parkland MH	White Rhododendron - Black Blueberry - Mountain Hemlock	not differentiated	
Coastal Douglas-fir (CDF)	Salal-Douglas-fir	Drier Maritime CDF	Salal - Oregon Grape - Douglas-fir	Nanaimo & Georgia Drier Maritime CDF	Salal - Oregon Grape - Douglas-fir on Dystric Brunisol with moder humus
		Wetter Maritime CDF	Oregon Grape - Salal - Douglas-fir	Nanaimo & Georgia Wetter Maritime CDF	Oregon Grape - Salal - Douglas-fir on Eluviated Dystric Brunisol with moder or friable mor humus

Coastal Western Hemlock (CWH)	Rhytidiadelphus - Western Hemlock	Drier Maritime CWH	Hylocomium - Rhytidiadelphus - Douglas-fir Western Hemlock	West Vancouver Island Drier Maritime CWH	Stokesiella - Rhytidiadelphus - Douglas-fir - Western Hemlock on Humo-Ferric Podzol with friable mor humus
				East Vancouver Island Drier Maritime CWH	Stokesiella - Hylocomium - Douglas-fir - Western Hemlock on Humo-Ferric Podzol with friable mor humus
				Pacific Ranges Drier Maritime CWH	Hylocomium - Rhytidiadelphus - Douglas-fir - Western Hemlock on Humo-Ferric Podzol with friable mor humus
		Wetter Maritime CWH	Rhytidiadelphus - Alaskan Blueberry - Amabilis Fir - Western Hemlock	Estevan Sub-montane Wetter Maritime CWH	Salal - Red Huckleberry & Alaskan Blueberry - Sitka Spruce & Amabilis Fir - Western Hemlock on Ferro-Humic Podzol with friable mor humus
				West Vancouver Island Submon-tane Wetter Maritime CWH	Rhytidiadelphus - Red Huckleberry & Alaskan Blueberry - Amabilis Fir - Western Hemlock on Ferro-Humic Podzol with friable mor humus
				West Vancouver Island Montane Wetter Maritime CWH	Rhytidiadelphus - Oval-leaved & Alaskan Blueberry - Amabilis Fir - Western Hemlock on Ferro-Humic Podzol with compacted mycelial mor humus
				East Vancouver Island Submon-tane Wetter Maritime CWH	Rhytidiadelphus - Red Huckleberry & Alaskan Blueberry - (Amabilis Fir) - Western Hemlock on Humo-Ferric Podzol with mycelial mor humus
				East Vancouver Island Montane Wetter Maritime CWH	Rhytidiopsis - Oval-leaved & Alaskan Blueberry - Amabilis Fir - Western Hemlock on Humo-Ferric Podzol with compacted mycelial mor humus
				Pacific Ranges Montane Wetter Maritime CWH	Rhytidiadelphus - Oval-leaved & Alaskan Blueberry - Amabilis Fir - Western Hemlock on Humo-Ferric or Ferro-Humic Podzols with compacted mycelial mor humus

Table 4. Characteristic Combinations of Species Employed as Accessory Characteristics for Biogeoclimatic Units - Nootka-Nanaimo Map Sheet

C H A R A C T E R I S T I C   C O M B I N A T I O N S   O F   S P E C I E S .

Zone	Combination of Species[1]	Subzone	Combination of Species[1]	Variant	Combination of Species[1]
Alpine Tundra (AT)	*Cassiope mertensiana* *Cassiope tetragona* *Loiseleuria procumbens* *Luetkea pectinata* *Lycopodium alpinum* *Lycopodium sitchense* *Phyllodoce empetriformis* *Cetraria subalpina*	Coastal AT	On this map sheet a single subzone is recognized in this zone. Thus, the characteristic combination of species for the subzone is identical to that for the zone.		
Mountain Hemlock (MH)	*Abies amabilis* *Chamaecyparis nootkatensis* *Tsuga mertensiana* *Menziesia ferruginea* *(Vaccinium membranaceum)* [2] *Athyrium distentifolium* * [2] *(Lycopodium annotinum)* *(Polystichum lonchitis*)* *(Rubus pedatus)* *Orthocaulis floerkii* *Pseudoleskea baileyi* *Rhytidiopsis robusta*	Maritime Forested MH	*(Tsuga heterophylla)* *Vaccinium alaskaense* *(Blechnum spicant)* *(Clintonia uniflora)* *(Streptopus streptopoides)* *(Tiarella unifoliata*)* *Dicranum pallidisetum* *Rhizomnium nudum**	West Vancouver Island Maritime Forested MH	*Cladothamus pyroliflorus* *Caltha biflora** *(Coptis asplenifolia)* *(Streptopus roseus*)*
				East Vancouver Island Maritime Forested MH	*(Abies lasiocarpa*)* *(Pinus monticola)* *(Gaultheria ovatifolia*)* *(Rhododendron albiflorum*)* *(Vaccinium ovalifolium)*
				Pacific Ranges Maritime Forested MH	*Cladothamnus pyroliflorus* *(Lonicera utahensis*)* *(Rhododendron albiflorum*)* *(Caltha biflora*)*
		Parkland MH	*(Cassiope mertensiana)* *Gaultheria humifusa* *(Phyllodoce empetriformis)* *(Rhododendron albiflorum)* *Vaccinium deliciosum* *Carex nigricans** *Luetkea pectinata*	No variant is recognized in this subzone	
Coastal Douglas-fir (CDF)	*Abies grandis* *Arbutus menziesii* *Cornus nuttallii* *Pseudotsuga menziesii* *(Amelanchier alnifolia)* *(Apocynum androsaemifolium*)* *Arctostaphylos columbiana** *Gaultheria shallon* *Holodiscus discolor* *(Juniperus scopulorum*)* *Mahonia aquifolium** *Mahonia nervosa* *Oemleria cerasiformis** *Ribes lobbii* *(Rosa gymnocarpa)* *Symphoricarpos mollis* *(Vaccinium parvifolium)* *Aira caryophyllea** *Allotropa virgata* *Boschniakia hookeri* *Campanula scouleri* *Delphinium menziesii** *Moehringia macrophylla* *Navarretia squarrosa** *Plectritis congesta** *Rhytidiadelphus triquetrus* *Stokesiella oregana* *Trachybryum megaptilum*	Drier Maritime CDF	*Quercus garryana** *Lonicera hispidula* *Brodiaea coronaria** *Bromus carinatus* *Camassia leichtlinii** *Camassia quamash** *Carex pensylvanica** *Claytonia perfoliata* *Collinsia grandiflora* *Danthonia californica** *Dodecatheon hendersonii* *Erythronium oregonum* *Galium aparine* *Lomatium utriculatum** *Mimulus alsinoides** *Nemophila parviflora* *Poa bulbosa** *Sanicula crassicaulis* *Sisyrinchium douglasii** *Vulpia microstachys**	Nanaimo & Georgia Drier Maritime CDF	A singla variant is recognized in each subzone of the CDF zone on this map sheet. Thus, characteristic combinations of species for the variants are identical to those for the subzones
		Wetter Maritime CDF	*(Tsuga heterophylla)* *(Paxistima myrsinites)* *(Carex rossii*)* *(Chimaphila menziesii)* *Hemitomes congestum* *(Hypopitys monotropa)* *(Listera cordata)* *(Pyrola picta)* *(Smilacina stellata)*	Nanaimo & Georgia Wetter Maritime CDF	

Coastal Western Hemlock (CWH)	*Tsuga heterophylla* *(Menziesia ferruginea)* *(Oplopanax horridus*)* *Blechnum spicant* *Cornus canadensis* *Dryopteris austriaca* *Bazzania ambigua* *Bazzania denudata* *Bazzania tricrenata* *Isopterygium elegans* *Isothecium spiculiferum* *Plagiothecium undulatum* *Rhytidiadelphus loreus* *Scapania bolanderi* *Lobaria oregana* *Usnea mollis*	Drier Maritime CWH	*(Acer macrophyllum)* *(Arbutus menziesii*)* *(Cornus nuttallii*)* *(Pseudotsuga menziesii)* *Gaultheria shallon* *Mahonia nervosa* *(Vaccinium parvifolium)* *Achlys triphylla* *(Stokesiella oregana)*	West Vancouver Island Drier Maritime CWH	*(Abies amabilis*)* *(Abies grandis*)* *(Vaccinium alaskaense)* *(Vaccinium ovalifolium)* *Disporum smithii*
				East Vancouver Island Drier Maritime CWH	*(Amelanchier alnifolia)* *(Mahonia aquifolium*)* *(Paxistima myrsinites*)* *(Rosa gymnocarpa)* *Boschniakia hookeri** *(Trachybryum megaptilum)*
				Pacific Ranges Drier Maritime CWH	This variant lacks a distinct characteristic combination of species. Thus, it is characterized by the subzonal combination of species and the absence of the species that occur in other variants in the subzone
		Wetter Maritime CWH	*Abies amabilis* *Vaccinium alaskaense* *(Vaccinium ovalifolium)* *(Clintonia uniflora)* *(Rubus pedatus)* *(Pilophoron clavatus*)*	Estevan Submontane Wetter Maritime CWH	*Picea sitchensis* *(Pinus contorta*)* *(Gaultheria shallon)* *Vaccinium ovatum* *(Vaccinium parvifolium)* *(Calamagrostis nutkaensis)* *(Polypodium scouleri)* *Usnea longissima*
				West Vancouver Island Submontane Wetter Maritime CWH	*(Picea sitchensis)* *(Gaultheria shallon)* *(Vaccinium parvifolium)* *Hookeria lucens* *Mylia taylori* *(Rhizomnium glabrescens)*
				West Vancouver Island Montane Wetter Maritime CWH	*Chamaecyparis nootkatensis* *(Tsuga mertensiana*)* *(Caltha biflora*)* *(Streptopus roseus*)* *(Streptopus streptopoides)* *(Tiarella unifoliata*)* *(Rhizomnium nudum*)* *(Rhytidiopsis robusta*)*
				East Vancouver Island Submontane Wetter Maritime CWH	*(Pinus monticola)* *(Pseudotsuga menziesii)* *(Gaultheria shallon)* *(Mahonia nervosa)* *(Rosa gymnocarpa*)* *(Vaccinium parvifolium)* *(Rhytdidiopsis robusta)*
				East Vancouver Island Montane Wetter Maritime CWH	*(Abies lasiocarpa*)* *Chamaecyparis nootkatensis* *(Pinus monticola*)* *(Pseudotsuga menziesii)* *(Gaultheria ovatifolia*)* *(Vaccinium membranaceum)* *Dicranum pallidisetum* *(Rhytidiopsis robusta)*
				Pacific Ranges Montane Wetter Maritime CWH	*Chamaecyparis nootkatensis* *(Tsuga mertensiana*)* *(Sorbus sitchensis)* *(Vaccinium membranaceum*)* *(Rhizomnium nudum*)* *(Rhytidiopsis robusta)*

[1] The selected plant species are arranged alphabetically in the order of trees, shrubs, herbs, mosses-liverworts, and lichens. The characteristic combination of species for a variant includes its combination of species plus those for the subzone and zone to which it belongs; similarly, the characteristic combination of species for a subzone includes it combination of species plus those for the zone to which it belongs. Within a zone, the combinations of species for subzones can be considered as differentiating combinations of species; and within a subzone the combinations of species for variants can be considered as differentiating combinations of species.

[2] Plant species listed in parenthesis are less characteristic or occur less frequently in a particular combination of species, however, they may be more characteristic of another (adjacent) biogeoclimatic unit. Plant species marked with an asterisk occur on drier, wetter, poorer, or richer soils than those of zonal ecosystems.

TABLE 5. Means and Standard Deviations of Differentiating Climatic Characteristics of the CDF, CWH, and MH zones in Coastal Southwestern British Columbia

	Biogeoclimatic Zones		
	CDF	CWH	MH
Number of climatic data sets	32	79	27
Climate (Köppen/Trewartha)	Csb	Cfb	Dfc
Discriminated parameters			
Actual evapotranspiration: May (mm)	97 (5.1)	85 (5.7)	26 (19.1)
Water deficit (mm)	222 (69.2)	41 (54.6)	19 (36.9)
Actual evapotranspiration: June (mm)	76 (25.6)	97 (6.6)	76 (23.2)
Number of months with mean temperature below 0°C	0.0 (0.0)	0.5 (1.06)	4.7 (0.72)
Actual evapotranspiration: July (mm)	27 (12.4)	92 (27.6)	98 (18.3)
Number of months with water deficit	3.7 (0.74)	1.0 (1.13)	0.6 (0.84)
Actual evapotranspiration: September (mm)	46 (7.0)	49 (5.1)	44 (4.7)
Water surplus (mm)	678 (218.8)	2235 (968.3)	3028 (1271.4)
Actual evapotranspiration: August (mm)	34 (11.1)	67 (17.1)	69 (12.4)
Actual evapotranspiration: April (mm)	56 (4.2)	40 (13.1)	0 (0.0)
Actual evapotranspiration: March (mm)	26 (3.2)	14 (11.1)	0 (0.0)
Maximum snow depth (cm)	5.9 (9.45)	59.5 (64.22)	294.2 (110.04)
Number of months with mean temperature over 10°C	5.4 (0.66)	4.0 (1.69)	0.8 (0.92)
Mean annual temperature (°C)	9.5 (0.72)	7.5 (1.83)	2.7 (0.84)

TABLE 6. An Example of Classification Functions Determined in the Discriminant Analysis

		Biogeoclimatic Zones		
		CDF	CWH	MH
Constant		-222	-192	-317
Parameter				
Actual evapotranspiration: May (mm)	x_1	-0.236	-0.524	-3.742
Water deficit (mm)	x_2	0.412	0.476	1.342
Actual evapotranspiration: June (mm)	x_3	0.596	0.791	2.875
Number of months with mean temperature below 0°C	x_4	19.7	20.0	35.9
Actual evapotranspiration: July (mm)	x_5	0.0531	0.281	0.921
Number of months with water deficit	x_6	15.23	10.93	1.17
Actual evapotranspiration: September (mm)	x_7	1.011	0.937	-0.722
Water surplus (mm)	x_8	0.0095	0.0110	0.0211
Actual evapotranspiration: August (mm)	x_9	1.69	1.52	2.06
Actual evapotranspiration: April (mm)	x_{10}	1.90	1.59	0.31
Actual evapotranspiration: March (mm)	x_{11}	-0.920	-0.997	0.574
Maximum snow depth (cm)	x_{12}	0.0592	0.0159	-0.0117
Number of months with mean temperature over 10°C	x_{13}	-20.4	-17.7	-25.0
Mean annual temperature (°C)	x_{14}	19.6	17.5	28.1

Using the above tables the functions are:

$$Y_{CDF} = -222 - 0.236(x_1) + 0.412(x_2) + 0.596(x_3) \; ..... + 19.6(x_{14});$$

$$Y_{CWH} = -192 - 0.524(x_1) + 0.476(x_2) + 0.791(x_3) \; ..... + 17.5(x_{14});$$

$$Y_{MH} + -317 - 3.742(x_1) + 1.342(x_2) + 2.875(x_3) \; ..... + 28.1(x_{14}).$$

The data are classified to the group with the largest Y.

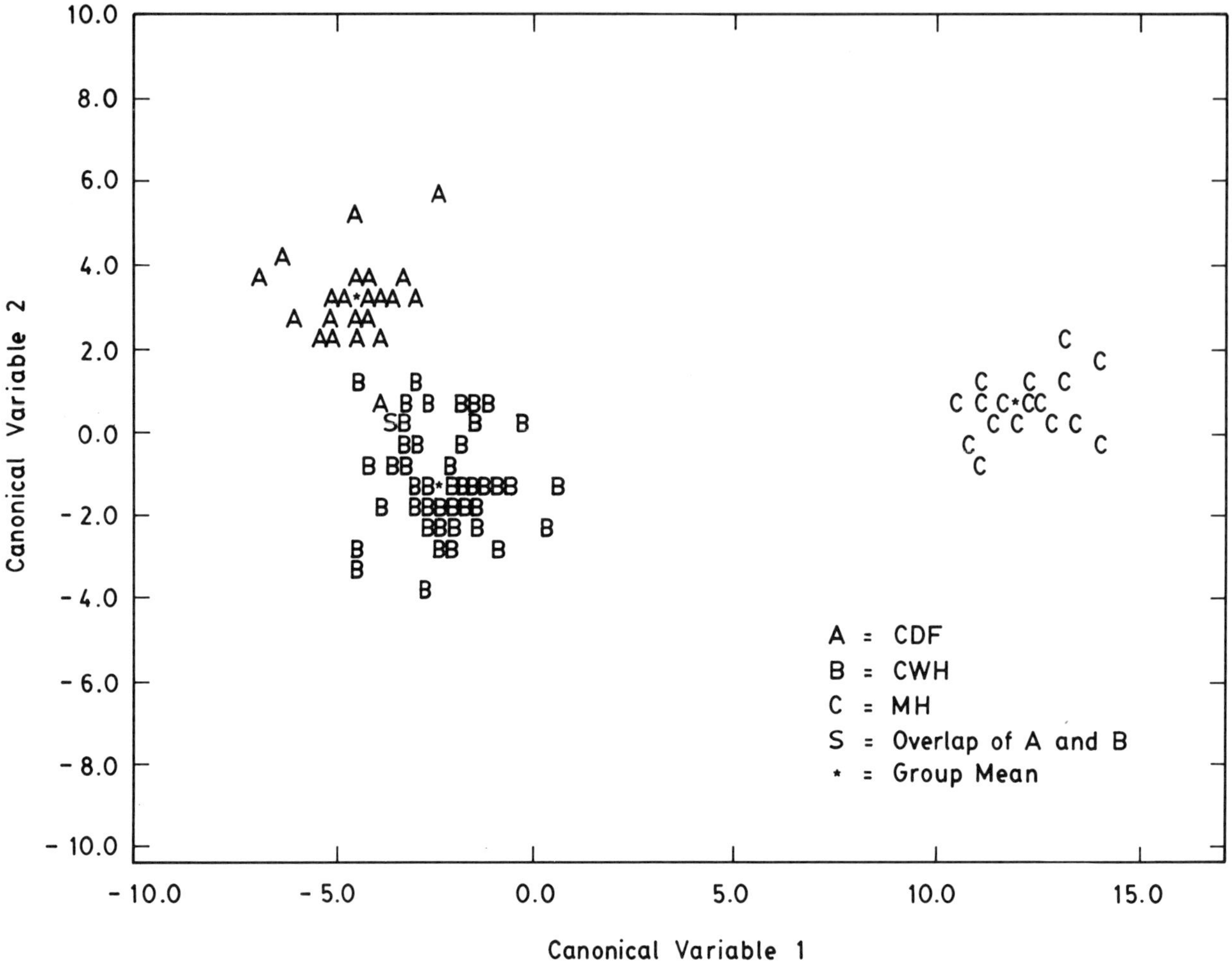

Figure 2. The plot of canonical variables for climatic observations which characterize the CDF, CWH, and MH zones.

there is virtually no plant growth in November, December, and January in any zone. Growth increases rapidly in February in the CDF and CWH zones,but a similar effect is not seen until early May in the lower parts of the MH Zone. In late May or early June, plant growth in the CDF Zone slows down until July due to an increasing water deficit. There is a slight increase in August and September due to moisture recharge. In contrast, the curves for the CWH and MH zones indicate practically no water deficit.

By inference from actual evapotranspiration over a year, plant growth appears to be greatest in the CWH Zone (463 mm), less in the CDF Zone (386 mm), and least in the MH Zone (322 mm). This sequence corresponds to a gross forest productivity rating given by Krajina (1965). The CDF Zone shows the highest PET (606 mm), followed by the CWH Zone (503 mm), and the MH Zone (341 mm). However, due to low summer precipitation resulting in a long soil

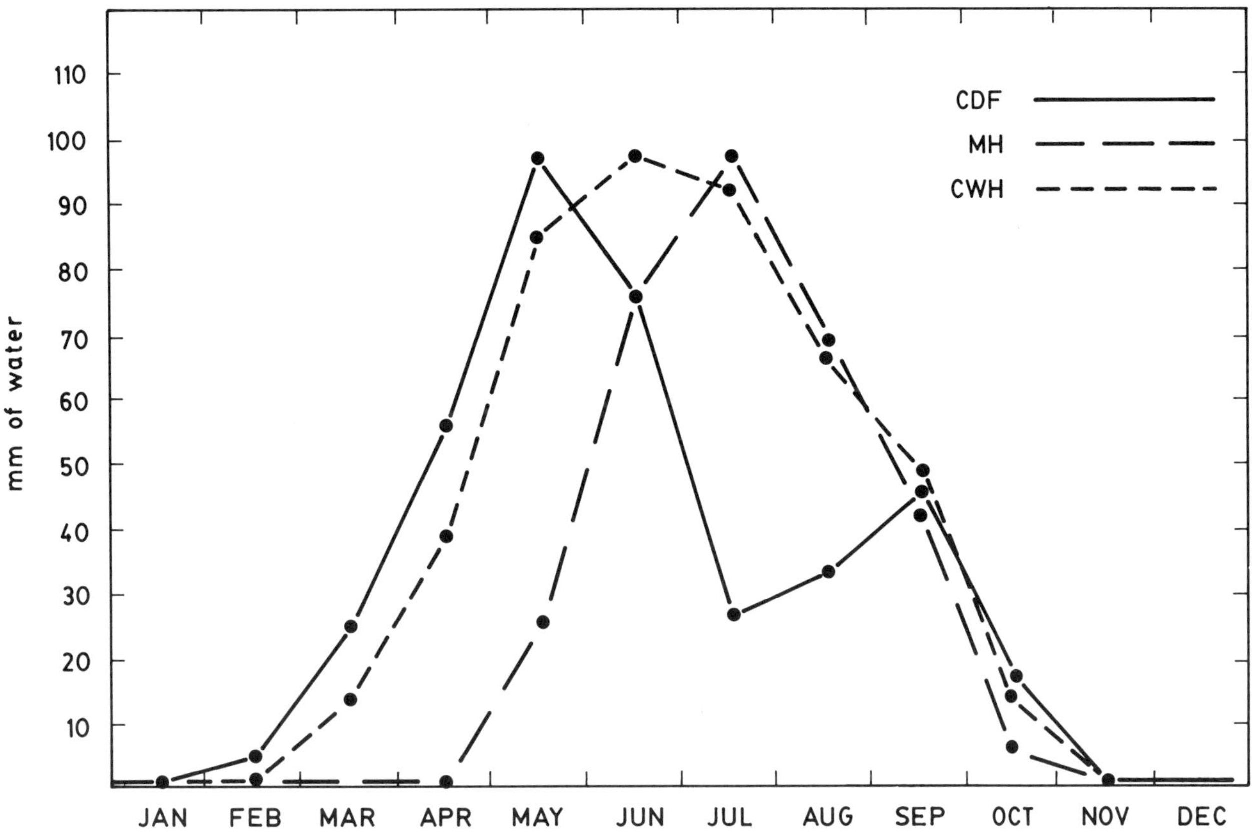

Figure 3. Calculated actual evapotranspiration curves for the CDF, CWH, and MH zones in coastal southwestern British Columbia.

moisture deficiency, only 64% of the potential evapotranspiration in the CDF Zone occurs as actual evapotranspiration. Thus, in this zone, water limits plant activity. For the CDF Zone, the difference between mean annual precipitation (1064 mm) and actual evapotranspiration is 678 mm. This is approximately the amount of water lost as runoff or groundwater during late fall and winter. Potential heat that cannot be used for plant growth in this zone because of lack of water is equivalent to that required to evaporate 220 mm of water. Therefore, soil moisture conservation must be the overall forest management objective in this zone (the only exceptions would be those ecosystems with additional inputs of water as seepage or groundwater). In the humid climates of the CWH and MH zones, heat alone limits vegetation growth since precipitation is more than adequate to cover potential evapotranspiration. Thus, lack of heat limits plant growth in winter, spring, and fall in the three zones and insufficient water in summer limits it in the CDF Zone.

Alpine Tundra Zone

The AT Zone includes non-forested ecosystems of high mountains; in southwestern British Columbia it occurs above 1500 m on their windward side and above 1800 m on their leeward side. Due to the variability of climate in the mountains, it may occasionally occur at lower elevations. The main differentiating climatic characteristic is the mean monthly temperature of the warmest month, which is between $0^{o}C$ and $10^{o}C$. The number of frost-free days varies from 25 to 105, but frost may occur at any time at higher elevations. More than 70% of the precipitation falls as snow.

Vegetation consists of herbs, mosses, liverworts, lichens, and very low shrubs. Except for some dwarfed or stunted conifers and deciduous woody plants at the lower elevational limits, no trees grow in this zone. Soils develop at a very slow rate, mainly when associated with vegetation cover and stable materials. The soils on zonal sites are usually shallow Regosol or Ferro-Humic Podzol with a high content of coarse fragments.

Within the zone, a Coastal AT Subzone is recognized. Particularly closer to the ocean, it is characterized by a heavy, wet snow cover of long duration. The existence of only two complete climatic data sets for the zone ruled out the discriminant analysis of climatic data. Fortunately, classification beyond the subzone category is not urgent since the priority for forest resource management is low.

On the Nootka-Nanaimo map sheet, the subzone is limited to a few high peaks in Strathcona Provincial Park. Recent glaciation, extreme physiography, and degradation processes severely limit the extent of alpine ecosystems in most Vancouver Island mountain ranges (Figure 4.) Much of the area is discontinuously covered with exposed bedrock, snow, ice, and unstable material. There are a few favourable habitats (*e.g.* where snow does not accumulate in great quantities and, therefore, melts earlier) allowing the development of floristically diversified plant communities. The biogeocoenotic type, generalizing zonal ecosystems is: Pink and Mertens' Mountain Heather on shallow, skeletal Regosol with root mor humus (Figures 5a and b). Mountain hemlock, subalpine fir, and whitebark pine may be found sporadically in stunted form at the lowest elevations of the subzone. Due to the extremely low rate of organic production and high sensitivity of alpine ecosystems to disturbance, management goals for the Coastal AT Subzone should be limited to conservation. This subzone has a potential for recreational activities such as skiing, hiking, climbing, and outdoor education. Last but not least, it serves as an important watershed area.

Figure 4. Recent glaciation, extreme physiography, and degradation processes limit development of alpine ecosystems in the majority of Vancouver Island mountain ranges (The Red Pillar; 2032 m).

Figures 5a and 5b. Mertens' mountain heather (*Cassiope mertensiana*) (left) and pink mountain heather (*Phyllodoce empetriformis*) (right) form the major component of the dense lower shrub layer in plant communities of zonal ecosystems in the Coastal AT Subzone.

Mountain Hemlock Zone

Maritime parts of the MH Zone in southwestern British Columbia are confined to elevations from 900 to 1500 m on the windward side, and from 1100 to 1800 m on the leeward side of mountains. These elevation limits decrease with increasing latitude. The most common trees are mountain hemlock (the climatic climax species in the zone), amabilis fir, and yellow-cedar. These species depend on unfrozen ground below snow (Krajina 1969, Brooke *et al*. 1970). Mountain hemlock and amabilis fir regenerate in abundance under the canopy of forest stands of many ecosystems, although advance regeneration of the former species is found less frequently in fully stocked and dense stands. Mountain hemlock eventually becomes established in the advanced stages of stand development after openings of various sizes have occurred in the forest canopy. At the lowest elevations of the zone, western hemlock, western redcedar, and rarely Douglas-fir and Sitka spruce may occur; at the highest elevations stunted subalpine fir and whitebark pine may be found.

Ecosystem development is greatly influenced by the cold climate. The deep winter snow is slow to disappear, and a short vegetative season is the result. The predominance of shrubs (*Vaccinium alaskaense, Vaccinium deliciosum, Vaccinium membranaceum,* and *Vaccinium ovalifolium*) and bryophytes (*Orthocaulis floerkii, Pseudoleskea baileyi,* and *Rhytidiopsis robusta*), along with the low presence of herbs and high species significance of advance regeneration of amabilis fir and mountain hemlock are characteristic floristic features of zonal ecosystems. The accumulation of acid decomposition products on the forest floor compacted by snow (mycelial mor humus formation), gleization (moist to saturated state of soils throughout the year), leaching, eluviation, and illuviation, are the characteristic soil processes (Krajina 1959, 1965, 1969, 1978; Brooke *et al*. 1970). Although the period in which soils have developed is comparatively short, they have generally acquired strong morphological characteristics. The low temperature and high moisture content in most soils promotes slow litter decomposition. Thus, they often have a high organic matter content (Brooke *et al*. 1970). Roots of mountain hemlock and, to a variable degree, of amabilis fir are largely confined to the forest floor, indicating its essential role in regeneration and growth.

In the Vancouver Forest Region, the zone is subdivided into Maritime Forested, Maritime Parkland, Subcontinental Forested (under study), and Subcontinental Parkland (under study) subzones. The climatic data of Maritime Forested and Parkland subzones were successfully separated in the analysis (96.3% of the individual observations were identically classified). A set of seven variables discriminated the observations. These are listed in order of importance in Table 7. Canonical variables

Table 7. Means and Standard Deviations of Differentiating Climatic Characteristics of the Maritime Forested and Parkland MH subzones

	Biogeoclimatic Subzones	
	Maritime Forested MH	Maritime Parkland MH
Number of climatic data sets	18	9
Climate (Köppen/Trewartha)	milder Dfc	colder Dfc
Discriminated Parameters		
Actual evapotranspiration May (mm)	40 (2.6)	0 (0.0)
Mean radiation during growing season (Ly)	40100 (1780)	39500 (1590)
Number of months with mean temperature over 10°C	1.1 (1.00)	0.3 (0.50)
Potential evapotranspiration (mm)	376 (26.5)	272 (21.8)
Mean precipitation of the driest month (mm)	115 (168.0)	76 (45.8)
Number of months with snow	7.3 (0.89)	8.4 (0.53)
Actual evapotranspiration: October (mm)	11 (4.3)	0 (0.0)

indicated a low disperison of observations around the categories and hence, a relatively distinct and abrupt boundary between the subzones. Calculation of actual evapotranspiration for subalpine ecosystems is difficult; consequently, the comparison of values between regional climates is limited to generalities. The difficulties arise due to the additional soil moisture provided by melting snow and the sudden vegetative activity following its disappearance. Until subalpine trees project above the winter snow level, they are clearly incapable of organic matter production (Brooke *et al.* 1970). Since snow depth and duration have a critical influence on plant growth, calculations of actual evapotranspiration were based on an estimated growing season of four months for the Maritime Parkland and six months for the Maritime Forested MH Subzone (Figure 6). Lack of heat combined with large quantities of snow limits plant activity of subalpine ecosystems. The slightly warmer climate, longer growing

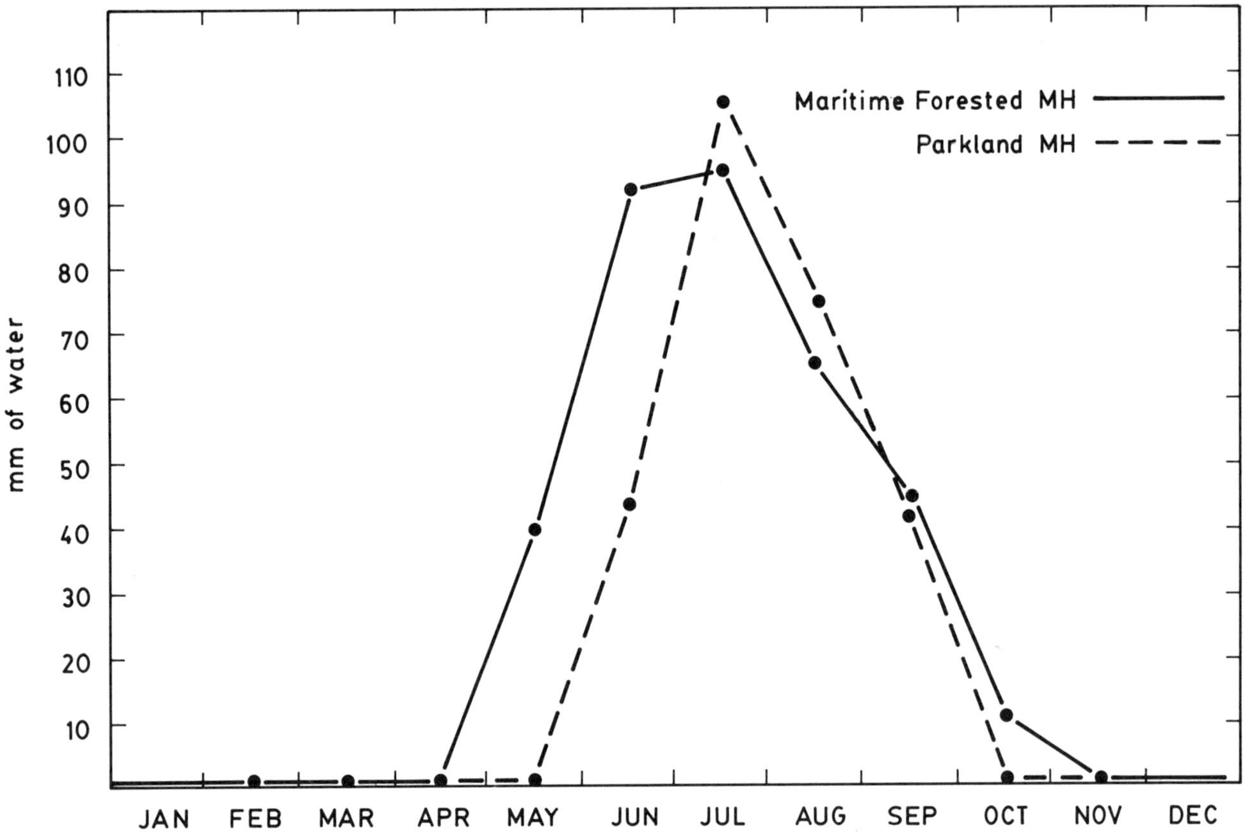

Figure 6. Calculated actual evapotranspiration curves for the Maritime Forested and Maritime Parkland MH subzones.

season, and shallower snow cover in the Maritime Forested MH Subzone result in increased plant activity. The potential evapotranspiration of this subzone is 376 mm, while for the Maritime Parkland MH Subzone it is 272 mm. With the exception of the AT Zone, the latter has the lowest potential evapotranspiration of the biogeoclimatic zones.

Only a selected number of ecosystems in the Maritime Forested MH subzone are suitable for wood production if a form of appropriate silvicultural management is used. Those remaining serve as watersheds and for recreation. Where harvesting is carried out, retention of natural regeneration (advance or subsequent to harvesting) and avoidance of broadcast burning are the main silvicultural considerations. Planting should be limited to a supporting role. Site-specific regulation of the tree species composition and intermediate cutting will be necessary if very dense stands are to be prevented. Progressive clearcutting (emphasizing shape rather than size), medium-sized cutting blocks, and a cutting sequence and total area related to the regeneration of mountain hemlock, amabilis fir, and yellow-cedar, are proposed as the major features of an adequate silvicultural system in the Maritime Forested MH Subzone.

The Maritime Forested MH Subzone is subdivided into three variants. No subdivision of the Maritime Parkland MH Subzone has been attempted, although parallel variants to those in the forested subzone should exist. In addition, an area of subalpine ecosystems in the northern-most coastal part of the region (Kingcome River, Wakeman River, Seymour River) could possibly be distinguished as a Maritime Northern MH Subzone.

Maritime Forested MH Subzone

The upper elevation limit of this subzone occurs at 1320 m on the windward side and at 1370 m on the leeward side of mountains. It is subdivided into West Vancouver Island, East Vancouver Island, and Pacific Ranges variants. The Pacific Ranges Variant was not included in the analysis due to the lack of representative climatic observations for most of the area. Comparison of available vegetation and environmental data from the West Vancouver Island and Pacific Ranges variants shows many affinities, but it appears the former has a wetter climate.

The analysis confirmed that the two Vancouver Island variants could be differentiated in the climatic data (88.9% of the individual observations were correctly classified) (Table 8). A combination of eight variables discriminated the observations, delineating a milder and wetter climate on the windward side and a drier and slightly subcontinental climate on the leeward side of Vancouver Island. As a result, plant growth, as indexed by actual evapotranspiration, is slightly more intense during the summer on the windward side (Figure 7). The climatic difference is reflected in the upward shift of approximately 200 m in the break between the CWH and MH zones, the altitudinal limits of tree species, the patterns of ecosystems, and some details of floristic composition.

Boundaries of the Maritime Forested MH Subzone are easily recognized since regional climatic gradients, occurring along steep altitudinal gradients, have a profound influence on ecosystem composition and structure. Presence of mountain hemlock in tree form and continuity of forest cover are the most characteristic features of this subzone. The existence of a greater proportion

Table 8. Means and Standard Deviations of Differentiating Climatic Characteristics of the West and East Vancouver Island Maritime Forested MH Variants

	Biogeoclimatic Variants	
	West Vancouver Island Maritime Forested MH	East Vancouver Island Maritime Forested MH
Number of climatic data sets	9	9
Discriminated Parameters		
Mean temperature of the warmest month (°C)	8.2 (0.89)	11.6 (0.60)
Mean precipitation of the driest month (mm)	186 (220.2)	44 (8.9)
Number of months with water deficit	0.0 (0.0)	1.6 (0.73)
Mean annual temperature (°C)	3.0 (0.27)	3.3 (0.61)
Mean precipitation April-September (mm)	1063 (165.2)	460 (127.0)
Actual evapotranspiration: June (mm)	89 (4.9)	94 (3.9)
Actual evapotranspiration: September (mm)	42 (0.9)	48 (4.3)
Index of continentality	2 (3.9)	16 (2.1)

of mountain hemlock than western hemlock in the stand composition is used to identify the lower limits. Local physiography is variable and strongly influences the duration and depth of snow. This affects the proportions of the above two species. It also causes considerable variation

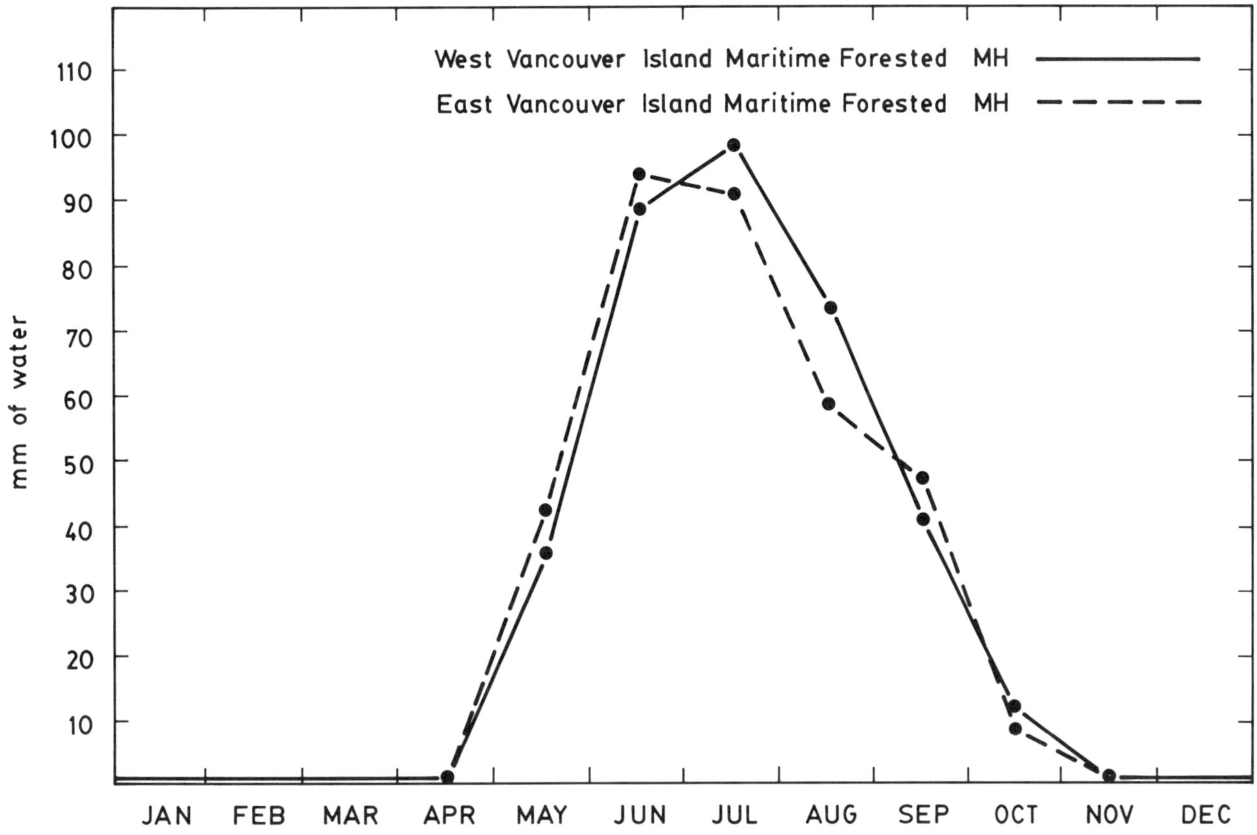

Figure 7. Calculated actual evapotranspiration curves for the West and East Vancouver Island Maritime Forested MH variants.

in the altitudinal limits of the lower boundary of the subzone. Snow duration in this subzone is usually at least six months. Nevertheless, complex forested ecosystems where amabilis fir and mountain hemlock are the most common trees are still established.

The lower limit of *Alectoria sarmentosa* and *Sphaerophorus globosus* on the trunks of trees indicates normal winter snow levels because they are not present below the snowline (Brooke *et al.* 1970). The distribution of these lichens gradually shifts upward on the trunks with increasing elevation and deeper snow.

West Vancouver Island Maritime Forested MH Variant

Ecosystems of this variant, likely the wettest of the Maritime Forested MH Subzone, have a small distribution, limited to the inward parts of the windward side of Vancouver Island where there is more continuity in the mountainous landscape. The biogeocoenotic type, generalizing zonal ecosystems is: Oval-leaved & Alaskan Blueberry - Amabilis fir - Mountain

Hemlock on Ferro-Humic Podzol with compacted mycelial mor humus (Figure 8). The maximum snow depth is estimated to be 281 cm (Appendix IIC). Dense stands of old growth amabilis fir and mountain hemlock sometimes resemble even-aged younger stands in structure; in this form they may have a high volume and good vigor. Advance regeneration of amabilis fir usually dominates shrub and herb layers, suppressing other understory vegetation (Figures 8 and 9).

Figure 8. A vigorous and fully stocked stand of old growth mountain hemlock and amabilis fir with abundant advance regeneration of amabilis fir in the lower shrub layer [4/C, Vaccinium (ovalifolium & alaskaense) - AA - TM; Townicut Mountain, Cowichan Lake, 1050 m].

The addition of *Cladothamus pyroliflorus*, *Caltha biflora*, *Coptis asplenifolia*, and *Streptopus roseus* to the characteristic combination of species for the Maritime Forested MH Subzone assists in the differentiation of the variant. A greater proportion of subhygric, hygric, and subhydric sites is found in this variant and also serves to characterize it.

Fall and winter gale winds are a major cause of disturbance on extensive areas of near-climax ecosystems. Following disturbance young, naturally regenerated stands of more or less even-aged amabilis fir and mountain hemlock exhibit very good growth and vigor.

Figure 9. A vigorous and fully stocked stand of old growth mountain hemlock, amabilis fir, and western hemlock with abundant shrub layers [3/C, Vaccinium (ovalifolium & alaskaense) - AA - TH & TM; Waring Peak, Gold River, 1005 m].

East Vancouver Island Maritime Forested MH Variant

This variant represents the driest part of the Maritime Forested MH Subzone. Its most common occurrences are at Forbidden Plateau, in the eastern parts of Strathcona Provincial Park, and at Beaufort Ridge. The biogeocoenotic type, generalizing zonal ecosystems is: Oval-leaved & Black Blueberry - Amabilis Fir - Mountain Hemlock on Ferro-Humic Podzol with compacted mycelial mor humus (Figures 10 and 11). The maximum snow depth is estimated to be 240 cm (Appendix IIC). A frequent occurrence of *Abies lasiocarpa, Pinus monticola, Gaultheria ovatifolia, Rhododendron albiflorum,* and *Vaccinium ovalifolium* and a shift toward a greater distribution of ecosystems with xeric, subxeric, and submesic moisture regimes are the accessory characteristics. At the lower elevational limits of the variant, the occasional occurrence of Douglas-fir and western white pine in stocked forest stands with trees of similar size, and appearing even-aged, indicates disturbance by fire. Due to the drier Dfc climate and fire history, amabilis fir is restricted to the lower tree layers in stands of mature trees where it has poor to fair vigor (Figure 12). Subalpine fir, commonly growing on very dry sites of rock outcrops, may grow successfully on less extreme sites with vigor comparable to that of amabilis fir (Figure 13).

Figure 10. A fully stocked stand of old growth mountain hemlock, amabilis fir, and yellow-cedar. In addition to a short growing season, forest productivity appears to be affected by a soil moisture deficit [4/C, Vaccinium (ovalifolium & membranaceum) - AA - TM; Castlecrag Mountain, Forbidden Plateau, 1040 m].

Figure 11. Distribution of trees in clusters is the characteristic structural feature of forest communities at the upper limits of the Maritime Forested MH Subzone [3/C, Vaccinium (ovalifolium & membranaceum) - (AA) - TM; Castlecrag Mountain, Forbidden Plateau, 1120 m].

Figure 12. An open stand of old growth mountain hemlock with amabilis fir and yellow-cedar in an advanced stage of individual and group reproduction. Extreme age variation is the characteristic feature of forest communities with a little disturbance [5/C, Rubus (pedatus) - Vaccinium (membranaceum) - CN & AA - TM; Ranald Creek, Upper Campbell Lake, 1070 m].

Figure 13. A frequent occurrence of subalpine fir (left) growing together with amabilis fir (right) on dry and rich sites is a characteristic feature of forest communities in the Maritime Forested MH Subzone on mountain slopes of eastern Vancouver Island [2/D, Vaccinium (membranaceum) & Rhododendron (albiflorum) - TM & CN - AL & AA; Tsable Lake, Beaufort Ridge, 1080 m].

Pacific Ranges Maritime Forested MH Variant

Ecosystems of this variant are confined the the windward side of the mainland coast. On the Nootka-Nanaimo map sheet they occur west of Powell River. The biogeocoenotic type, generalizing zonal ecosystems is: Alaskan & Black Blueberry - Amabilis Fir - Mountain Hemlock on Ferro-Humic Podzol with compacted mycelial mor humus. The approximate average snow depth is estimated at 300 cm (Brooke et al. 1970). The occurrence of *Cladothamnus pyroliflorus*, *Lonicera utahensis*, *Rhododendron albiflorum*, and *Caltha biflora* are differentiating species for the variant.

Maritime Parkland MH Subzone

This subzone lies between the Coastal AT and Maritime Forested MH Subzone. Both the upper and lower boundaries are quite irregular due to considerable variation in climate within short distances. In general, the limits are between 1300 and 1700 m.

A scattered tree cover and long duration of snow are the most obvious characteristics of this subzone. Trees grow largely on topographic prominences where snow duration averages about eight months, either as isolated individuals or often in small clumps. At the lower limit, the Maritime Parkland MH Subzone is coincident with the beginning of discontinuous forest cover. However, discontinuous forest cover, when associated with areas of bare rock, may occur in other subzones (Figure 14). The upper limit of this subzone is based on the presence of mountain hemlock in tree form. This differentiates it from the Coastal AT Subzone where mountain hemlock may occur but only as a krummholz growth form (Archer 1963).

The biogeocoenotic type, generalizing zonal ecosystems is: White Rhododendron - Black Blueberry - Mountain Hemlock on Folisol or shallow Ferro-Humic Podzol with compacted mycelial mor humus (Figures 15 and 16). Mountain hemlock grows mainly on raised mounds of organic matter some of which are more than 1 m deep. The heath-like or low shrub vegetation with few herbs (*Cassiope mertensiana, Gaultheria humifusa, Phyllodoce empetriformis, Rhododendron albiflorum, Vaccinium deliciosum, Carex nigricans*, and *Luetkea pectinata*) is affected by long snow duration and covers much of the remaining area. Subalpine fir is a frequent component of the Maritime Parkland MH Subzone, particularly on the mountain slopes of eastern Vancouver Island (Figure 17).

Figure 14. A discontinuous forest cover developed on severely glaciated ridges due to edaphic rather than climatic factors (Sebalhall Creek, Gold River, 1450 m).

Figure 15. Complex of non-forested [Phyllodoce (empetriformis) - Cassiope (mertensiana)] and forested [Rhododendron (albiflorum) - Vaccinium (membranaceum) - TM] ecosystems on the mountain slopes of western Vancouver Island (Tahsis, 1320 m).

Figure 16. A complex of non-forested [Phyllodoce (empetriformis) - Cassiope (mertensiana)] and forested [Rhododendron (albiflorum) - Vaccinium (membranaceum) - TM] ecosystems on the mountain slopes of eastern Vancouver Island (Forbidden Plateau, 1290 m).

Figure 17. A frequent occurrence of subalpine fir on dry and rich sites is the characteristic feature of forest communities in the Parkland MH Subzone on the mountain slopes of eastern Vancouver Island (Buttle Lake, 1400 m).

Coastal Douglas-fir Zone

This is the driest mesothermal zone of British Columbia (Table 5). Ecosystems of this zone occur along the eastern side of Vancouver Island, on the Gulf Islands and on the adjacent coastal mainland between 48° and 50° 20' N latitude. Elevation may range from sea level to 150 m in the north and to 450 m in the south. This zone extends south into Washington and Oregon.

Douglas-fir is the most common tree species in the composition of the forest cover. It can regenerate under the canopy of mature and partly open forest stands on most sites. Western redcedar, grand fir, Pacific madrone, and Garry oak may frequently accompany Douglas-fir, depending on the soil moisture and nutrient regimes of the sites. Presently, the tree species composition of the forest cover may vary considerably as a result of widespread disturbance. Predominance of smaller shrubs (*Gaultheria shallon* and *Mahonia nervosa*), low presence of herbs and mosses, and the moderate shade tolerance of Douglas-fir are characteristic floristic features of zonal ecosystems. Moder to weak mor formation, melanization, weak laterization, and weak leaching are described by Krajina (1959, 1965, 1969, 1978) as the characteristic soil processes. The soils of zonal ecosystems are Dystric Brunisols with medium base saturation, grading with increasing precipitation into Humo-Ferric Podzols (usually without the development of an eluvial horizon) with moder or friable mor-humus. The high potential for organic production is limited by a summer water deficit. Thus, soil water conservation appears to be the most effective way of increasing forest productivity.

Although agriculture is possible on medium- to fine-textured soils, wood production still represents a good management option for many ecosystems. Douglas-fir, western redcedar, grand fir, and lodgepole pine are the most suitable species for second growth stands. In deciding whether to plant a species alone or in compatible combinations, the specific sites should be considered. Progressive clearcutting with cutting blocks of small to medium size and flexibility in shape to accommodate major environmental differences, followed by immediate planting of seedlings, preferably on shaded microsites, is the suggested silviculture system. Early control over the number of stems is necessary since Douglas-fir tends to form dense stands due to its shade tolerance. A simple shelterwood system using Douglas-fir, grand fir, and western redcedar may also be applied on selected sites.

Broadcast slash burning should be restricted to certain subhygric or hygric ecosystems and others where negative effects would be minimized while water conservation is increased. On many ecosystems slash burning can have undesirable effects, *e.g.* it can increase hydrophobicity, change albedo with resultant heat injury to seedlings, decrease insulating organic matter on the soil, decrease shaded microsites, and cause the loss of nutrients through leaching.

Three subzones are recognized in this zone. Of these, the Wetter Subcontinental CDF Subzone lies east of the mapped area. Both the Drier and Wetter Maritime CDF subzones are recognized on southeastern Vancouver Island. Here the boundary between them is as high as 150 m. Both subzones also occur on the southern mainland coast and further north where the elevation limit is lower.

The discriminant analysis of data confirms that the subdivision of the zone into Drier and Wetter Maritime CDF Subzones has a climatic basis. A close relationship between these subzones is indicated since only 87.5% of the individual observations were classified into the correct groups (Table 9). A set of four variables discriminated the observations. According to the analysis, the Vananda (Texada Island) and Parksville climatic stations are in the Wetter

Table 9. Means and Standard Deviations of Differentiating Climatic Characteristics of the Drier and Wetter Maritime CDF Subzones

	Biogeoclimatic Subzones	
	Drier Maritime CDF	Wetter Maritime CDF
Number of climatic data sets	15	17
Climate (Köppen/Trewartha)	drier Csb	wetter Csb
Discriminated Parameters		
Mean annual temperature (°C)	10.0 (0.40)	9.1 (0.72)
Mean precipitation April-September (mm)	214 (51.7)	279 (52.2)
Actual evapotranspiration: May (mm)	99 (4.7)	95 (4.7)
Number of months with snow	0.2 (0.41)	1.0 (1.00)

Maritime CDF Subzone. However, ground surveys of ecosystems in these areas confirmed the original classification into the drier subzone. This illustrates the weak differences between the subzones north of the 49th parallel.

The plot of actual evapotranspiration shows that there is little difference between the Drier and Wetter Maritime CDF subzones, particularly from January until May and again from September until December (Figure 18). Higher summer precipitation accounts for a higher actual evapotranspiration in the wetter than in the drier subzone (405 as opposed to 364 mm, Appendix IID). The highest potential evapotranspiration of all biogeoclimatic categories covered in this report (632 mm) was calculated for the drier subzone. Precipitation accounts for 58% of PET in the Drier Maritime CDF Subzone. In the Wetter Maritime CDF Subzone it rises to 69%. This is reflected in the difference between the actual evapotranspiration curves for May through September.

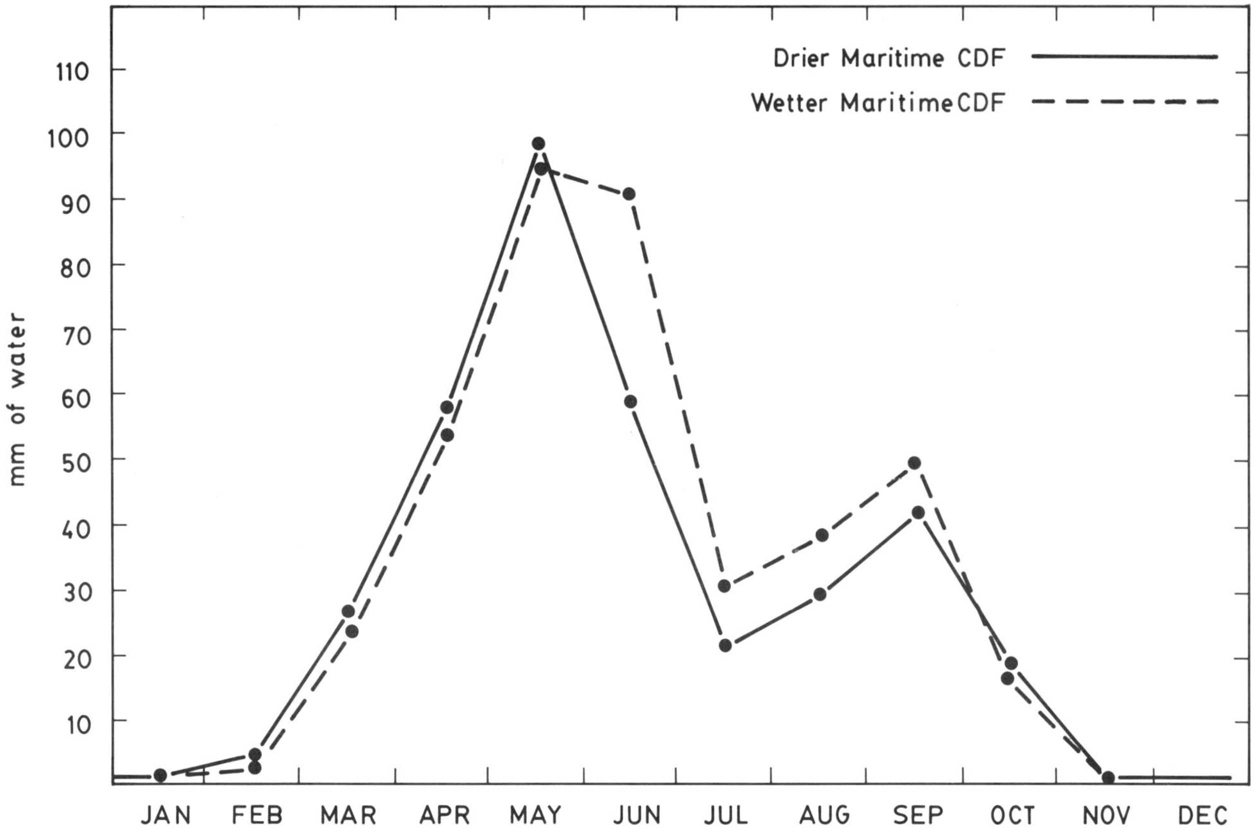

Figure 18. Calculated actual evapotranspiration for the Drier and Wetter Maritime CDF subzones.

Drier Maritime CDF Subzone

A single Nanaimo & Georgia Variant is distinguished in this subzone. Its ecosystems are distributed along the Strait of Georgia. They are characterized by a drier Csb climate. Areas other than those on southern Vancouver Island (*i.e.* on Saanich Peninsula and the Gulf Islands) are broadly transitional towards the Wetter Maritime CDF Subzone.

A higher shade tolerance of Douglas-fir (on dry sites it appears to be shade requiring) on the majority of sites is an important silvicultural characteristic of this dominant tree. Consequently, Douglas-fir may grow in all strata of mature forest stands. On dry sites, Douglas-fir is usually replaced by Garry oak, Pacific madrone, or lodgepole pine. Western hemlock is nearly absent with the exception of wet habitats where it may occasionally grow on the decaying wood of coniferous species.

The biogeocoenotic type, generalizing zonal ecosystems is : Salal - Oregon grape - Douglas-fir on Dystric Brunisol with moder humus. A high species significance of several grass species indicates melanization and moder formation.

The differentiating climatic characteristics of the Drier Maritime CDF Subzone (Table 9) also apply to its Nanaimo & Georgia Variant. Similarly Figure 18 depicts the variant's monthly actual evapotranspiration.

Present forest ecosystems are represented by stands of immature to mature second growth trees which have developed naturally after recent forest fires or early logging. The understory vegetation in the stands is developed well enough so that ecosystem identification is possible (Figures 19, 20, and 21).

Figure 19. A mixed stand of nearly mature Garry oak, bigleaf maple, Pacific madrone, western flowering dogwood, and Douglas-fir. [2/D, Bromus (vulgaris) - Mahonia (nervosa) - QG - PM; Yellow Point].

Figure 20. A dense stand of nearly mature lodgepole pine and Douglas-fir with sparse advance regeneration of Douglas-fir and hemlock. A component of western redcedar could improve the soil nutrient regime and suppress a well-developed brush layer of salal [4/B, Mahonia (nervosa) - Gaultheria (shallon) - (PC) - PM; Parksville].

Figure 21. A stand of nearly mature Douglas-fir with well developed brush layers (ocean spray and Oregon grape [4/D, Stokesiella (oregana) - Mahonia (nervosa) - PM; Yellow Point].

Wetter Maritime CDF Subzone

Two variants are distinguished in this subzone: Nanaimo & Georgia and Fraser. When compared to the Fraser Wetter Maritime CDF Variant, the Nanaimo & Georgia Variant is characterized by more months with a water deficit, lower mean radiation during the growing season, more months with snow, as well as six other discriminating parameters (Table 10).

Ecosystems of the Nanaimo & Georgia Wetter Maritime CDF Variant are adjacent to the Drier Maritime CDF Subzone in the rain shadow area along the Strait of Georgia. The Fraser Wetter Maritime CDF Variant does not occur in the area covered by the Nootka-Nanaimo map sheet. Thus, in the rest of the report a discussion of the subzone refers to the Nanaimo & Georgia Variant.

Mapping of the extent of the Wetter Maritime CDF Subzone generally is in agreement with the comparable subzone on the map of biogeoclimatic subzones of Vancouver Island and the adjacent mainland (MacMillan Bloedel 1974) with the exception of the area around Port Alberni. Only a small area west of Port Alberni (Lupsi Cupsi climatic station) has both a wetter Csb climate and ecosystem features characteristic of the Wetter Maritime CDF Subzone (Figure 22). Other areas of the Alberni Basin (Beaver Creek and Port Alberni climatic stations) are under the influence of a drier and warmer Cfb climate and have ecosystem features characteristic of the Drier Maritime CWH Subzone. Therefore, on the Nootka-Nanaimo map sheet the extent of the Wetter Maritime CDF Subzone is smaller than that of the comparable subzone on the MacMillan Bloedel map. The converse is true for the Drier Maritime CWH Subzone: It is greater in extent on the Nootka-Nanaimo map sheet.

Table 10. Means and Standard Deviations of Differentiating Climatic Characteristics of the Nanaimo & Georgia and Fraser Wetter Maritime CDF Variants

	Biogeoclimatic Variants	
	Nanaimo & Georgia Wetter Maritime CDF	Fraser Wetter Maritime CDF
Number of climatic data sets	11	6
Discriminated Parameters		
Number of months with water deficit	3.8 (0.40)	2.7 (0.52)
Mean radiation during growing season (Ly)	42600 (1240)	43800 (1510)
Number of months with snow	1.5 (0.93)	0.2 (0.41)
Actual evapotranspiration: May (mm)	94 (4.7)	98 (4.0)
Index of continentality	14 (2.6)	10 (5.1)
Frost-free period (days over 0^oC)	207 (18.3)	234 (23.0)
Actual evapotranspiration: February (mm)	4 (3.2)	4 (0.5)
Accumulated degree days over 5.6^oC	1534 (196.8)	1660 (140.8)
Actual evapotranspiration: April (mm)	53 (1.3)	56 (1.9)

Douglas-fir regenerates under the partly open or open canopy of mature stands (Figure 23). Garry-oak - common in the drier subzone - is, with the exception of rich sites in the Comox area and on Sumas Mountain (Krajina 1976), almost totally absent from the Wetter Maritime CDF Subzone. Western hemlock occurs here more frequently than in the Drier Maritime CDF Subzone but grows with poor or fair vigor. It is restricted to the lower tree or upper shrub layers since it requires shade in the CDF Zone. Western white pine and grand fir grow as scattered individuals with a good vigor on sites with richer nutrient regimes, suggesting a promising

Figure 22. A fragment of an open stand of old growth Douglas-fir with poorly growing advance regeneration of Douglas-fir in shrub layers [4/C, Gaultheria (shallon) - Mahonia (nervosa) - PM; Alberni, Lupsi Cupsi].

potential in managed stands. Sitka spruce, growing with low vigor, is limited to the seashore in the northern parts of the subzone. The biogeocoenotic type, generalizing zonal ecosystems is: Oregon grape - Salal - Douglas-fir on Eluviated Dystric Brunisol or incipient Humo-Ferric Podzol with moder or friable mor humus (Figures 23 and 24).

Figure 23. An open stand of mature Douglas-fir with advance regeneration of Douglas-fir (in openings) and some western hemlock and grand fir indicates a potential for a shelterwood system. [3/C, Mahonia (nervosa) - Gaultheria (shallon) - PM; Koksilah River].

Figure 24. A partly open stand of mature Douglas-fir with an increasing understory of salal and bracken in response to decreasing stocking [4/C, Mahonia (nervosa) - Gaultheria (shallon) - PM; Koksilah River].

Coastal Western Hemlock Zone

This is the wettest mesothermal zone of British Columbia and the largest in the Vancouver Forest Region. It covers much of Vancouver Island and the Coast Mountains. Upper elevations of the zone are 900 m on the windward side and 1100 m on the leeward side of mountains in southwestern British Columbia. Outside of the rain shadow area, it extends to sea level. Like the CDF Zone, the CWH Zone continues along the Pacific coast into Washington and Oregon.

Western hemlock is usually the most common species in the forest cover. It regenerates in abundance under the canopy of forest stands on zonal sites and elsewhere if there is enough accumulation of acid raw humus or decaying wood on the forest floor. If undisturbed, it has the greatest potential of all species to dominate the forest cover. Throughout the zone, Douglas-fir and western redcedar occur frequently, while amabilis fir and yellow-cedar are common only in the wetter CWH subzones. Their distribution is modified by disturbances in the past and the characteristics of the site, particularly its moisture and nutrient regime. The predominance of several moss species (*Hylocomium splendens, Rhytidiadelphus loreus,* and *Plagiothecium undulatum*) along with the low presence of herbs and a high species significance of western hemlock are the characteristic floristic features of zonal ecosystems. Accumulation of acid decomposition products on the forest floor (mor formation), leaching, eluviation, illuviation, and gleization are described by Krajina (1959, 1965, 1969, 1978) as the characteristic soil processes. The soils of zonal ecosystems are Humo-Ferric Podzols with mor humus grading, with increasing precipitation, into Ferro-Humic Podzols with mor humus. Accumulation of kaolinite and other weathering products from acidic parent materials such as granodiorites and, frequently, subsequent clay destruction by organic acids are slow. In contrast, the leaching potential in humid and particularly in perhumid climates is great. Therefore, particularly on nutritionally poor sandy-skeletal soils, conservation of nutrients in the vegetation or in the organic horizons under these climatic conditions helps to maintain the potential of the ecosystems for forest production.

Valentine and Lavkulich (1978) discussed Podzolic soils and their formation in this zone. They stated that the soils of many ecosystems lack an eluvial horizon, and explained that the addition of organic colloids to, and the rate of weathering of iron and aluminium in, this horizon are rapid enough to offset the heavy leaching. Frequently, a layer of dark-colored organic colloids marks the Ae horizon under moist field conditions. They also mentioned that the formation of Podzolic B horizons at higher elevations can happen in the soils of ecosystems which do not have the typically high rainfall, dense conifers, or acid bedrock, due to the release of large amounts of iron and, especially, aluminum from volcanic ash (which often occurs in significant quantities in the upper volume of the soils).

In the Drier Maritime CWH Subzone, Douglas-fir, grand fir, western white pine, and western redcedar; and in the Wetter Maritime CWH Subzone western hemlock, amabilis fir, yellow-cedar, and Sitka spruce have the highest potential for productivity compared to the other zones in which they grow. Therefore, the predominant use of many ecosystems for wood production is a definite management option. The most suitable major species are Douglas-fir, western hemlock, and amabilis fir; the suggested minor species are western redcedar, Sitka spruce, yellow-cedar, grand fir, western white pine, and lodgepole pine. Minor species may be promoted along with the major species in silviculturally compatible combinations according to the subzone and the characteristics of the site, particularly its soil moisture and nutrient regime. Progressive clear-cutting with medium-to-large cutting blocks, shaped to allow for major environmental differences, followed by immediate planting of species unable to regenerate naturally is a suitable silvicultural system for the CWH Zone. In addition, a simple shelterwood system for western hemlock and amabilis fir could be applied to selected ecosystems in the wetter subzones.

On those ecosystems where there are considerable quantities of nutrients present in the mineral component of the soil and where slash accumulations must be decreased for protection or silvicultural purposes, broadcast slashburning is recommended. For other ecosystems, the advantages of burning must be weighed against the disadvantages, and the appropriate decision must be made. One of these disadvantages arises because the already high potential for soil erosion due to the mountainous relief, high precipitation, and lithological discontinuities increases when the continuity of vegetative cover is interrupted over large areas and humus horizons are destroyed by slash fires.

The part of the CWH Zone that occurs in the Vancouver Forest Region is divided into five subzones, with the maritime CWH subzones being further subdivided into 10 variants. The subcontinental and north CWH subzones are under study as is the subzonal division of the northern parts of the CWH Zone in the Prince Rupert and Vancouver Forest Regions. Preliminary results have revealed significant ecological differences between the northern part of Vancouver Island (including the adjacent mainland) and the other parts of the Wetter Maritime

CWH Subzone in the south. Based on the large extent of the area involved, the magnitude of differences between the ecosystems in the north and those in the south, and the fact that drier (Alert Bay, Bonilla Island, Bull Harbor, Port Hardy, and Quatsino climatic stations) and wetter (Bella Bella, Holberg, Namu, and Ocean Falls climatic stations) variations of a mesothermal marine climate exist, a new subzone rather than a variant was recognized. The new subzone does not occur on the area covered by the Nootka-Nanaimo map sheet, but it is included in this report to portray a more complete image of the zone.

The climatic data from stations and prediction points in the Drier and Wetter Maritime (southern), and the newly-recognized Wetter Maritime Northern CWH subzones were used in the discriminant analysis. In it, 98.7% of the climatic stations were classified into the same subzones as were recognized on the ground. A large set of 14 climatic variables was selected for the classification functions (Table 11). The plot of canonical variables shows three distinct clusters of medium dispersion demonstrating the significant differentiation of climatic observations by the discriminant analysis (Figure 25). Despite the indication in this figure of a definite distinction between the southern and northern wetter maritime subzones, there is a rather broad transition on the ground.

Actual evapotranspiration curves for the subzones are similar (Figure 26). In essence, the lack of heat limits plant growth in all three subzones. Potential evapotranspiration decreases in the order from the Drier (556 mm), to the Wetter (489 mm), and to the Wetter Maritime Northern CWH Subzone (466 mm) (Appendix IIF). Actual evapotranspiration is highest in the Drier Maritime CWH Subzone (483 mm) and roughly equal in the Wetter Maritime (455 mm) and the Wetter Maritime Northern (460 mm) subzones. In fact, due to the small water deficit and long growing season, the Drier Maritime CWH Subzone has the highest actual evapotranspiration and hence the highest productivity potential among all biogeoclimatic subzones discussed.

Drier Maritime CWH Subzone

In this subzone, western hemlock grows with a high species significance in almost all layers of forest stands with the exception of some dry sites. However, Douglas-fir, a pioneer tree on the majority of sites, grows here with better vigor, particularly on nutrient rich soils. In contrast to the Wetter Maritime CDF Subzone, Douglas-fir is shade intolerant in this subzone and, except on dry sites, cannot regenerate under its canopy. These features, plus

Table 11. Means and Standard Deviations of Differentiating Climatic Characteristics for the Drier and Wetter Maritime CWH and Wetter Maritime Northern CWH Subzones

	Biogeoclimatic Subzones		
	Drier Maritime CWH	Wetter Maritime CWH	Northern Wetter Maritime CWH
Number of climatic data sets	20	50	9
Climate (Köppen/Trewartha)	drier Cfb	wetter Cfb/c	wetter Cfb
Discriminated Parameters			
Mean temperature of the warmest month (^{o}C)	17.3 (1.11)	12.9 (1.89)	14.5 (1.63)
Number of months with mean temperature over 10°C	5.4 (0.51)	3.4 (1.75)	4.4 (0.73)
Actual evapotranspiration: April (mm)	50 (2.5)	35 (14.2)	45 (3.5)
Actual evapotranspiration: February (mm)	3 (1.5)	2 (1.9)	1 (1.1)
Actual evapotranspiration: June (mm)	103 (4.8)	97 (4.8)	89 (8.5)
Actual evapotranspiration: August (mm)	59 (14.9)	69 (17.4)	76 (12.7)
Ratio actual to potential evapotranspiration (%)	87 (8.2)	93 (11.4)	99 (2.5)
Water surplus (mm)	1383 (275.7)	2535 (896.7)	2460 (1266.0)
Number of months with mean temperature below 0°C	0.0 (0.0)	0.8 (1.23)	0.0 (0.0)
Maximum snow depth (cm)	19 (17.1)	80 (70.9)	38 (39.7)
Number of months with water deficit	1.8 (0.72)	0.8 (1.18)	0.6 (0.88)
Actual evapotranspiration: July (mm)	90 (31.4)	92 (27.9)	96 (17.6)
Mean precipitation April-September (mm)	468 (68.2)	713 (261.8)	891 (466.4)
Actual evapotranspiration: April-September (mm)	440 (39.4)	426 (52.5)	431 (50.3)

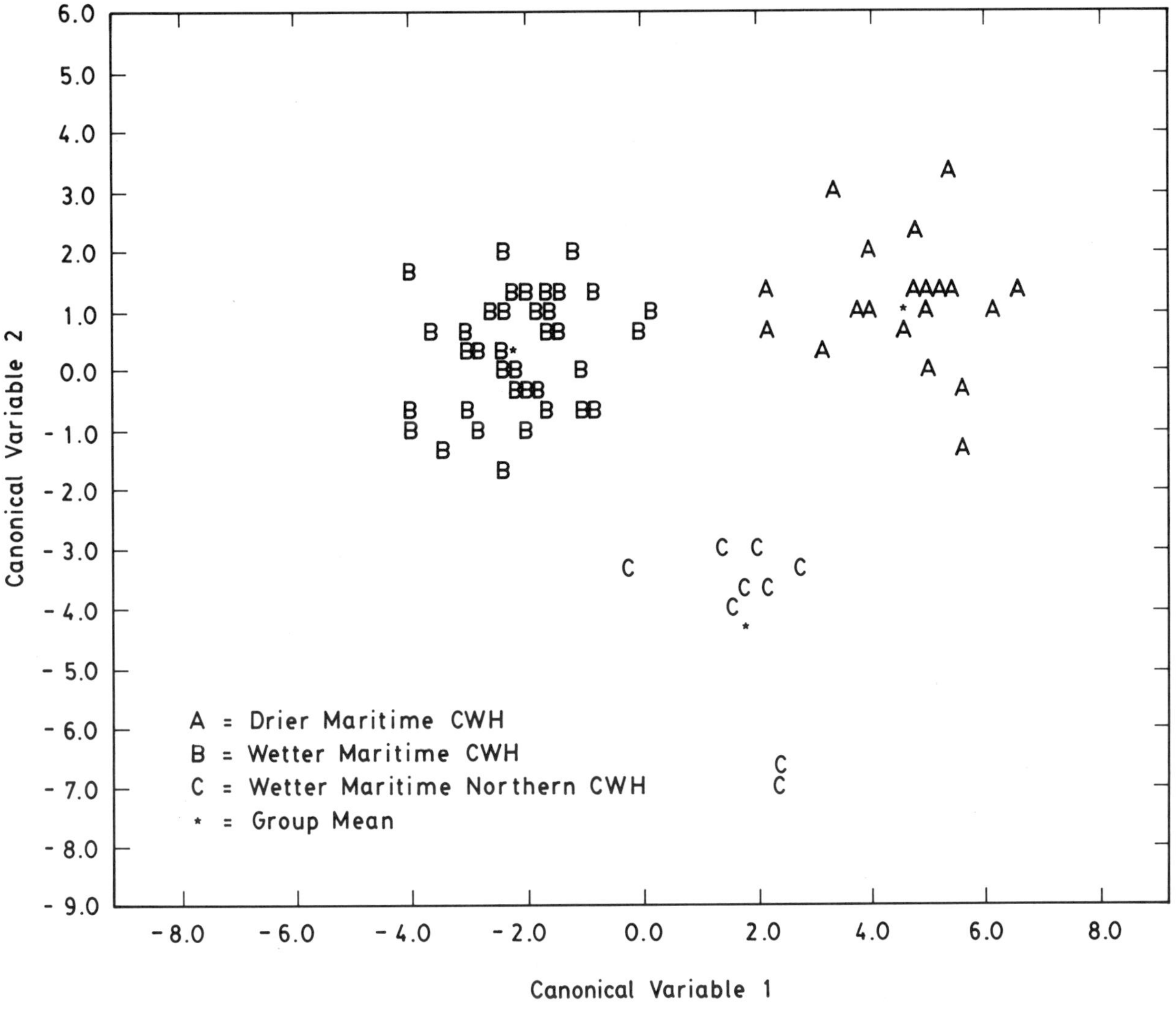

Figure 25. Plot of the first canonical variable against the second for individual climatic observations characteristic of the maritime CWH subzones.

the characteristic combination of species (Table 4) and, mainly the presence of zonal ecosystems generalized as the Hylocomium - Rhytidiadelphus - Douglas-fir - Western Hemlock biogeocoenotic association are most useful for ground identification. The upper limit of the subzone may be as high as 600 m on the windward and 700 m on the leeward side of mountains.

The subzone is divided into the West Vancouver Island, East Vancouver Island, and Pacific Ranges Drier Maritime CWH variants. The West Vancouver Island Drier Maritime CWH Variant could not be included in the analysis because there are no long term climatic data available for the area it covers. All indirect evidence indicates that the climate of this variant occupies the wetter part of the climatic spectrum characteristic of the subzone.

In the discriminant analysis, 100% of the climatic stations were classified into the groups corresponding to the two remaining variants. Two variables were sufficient to

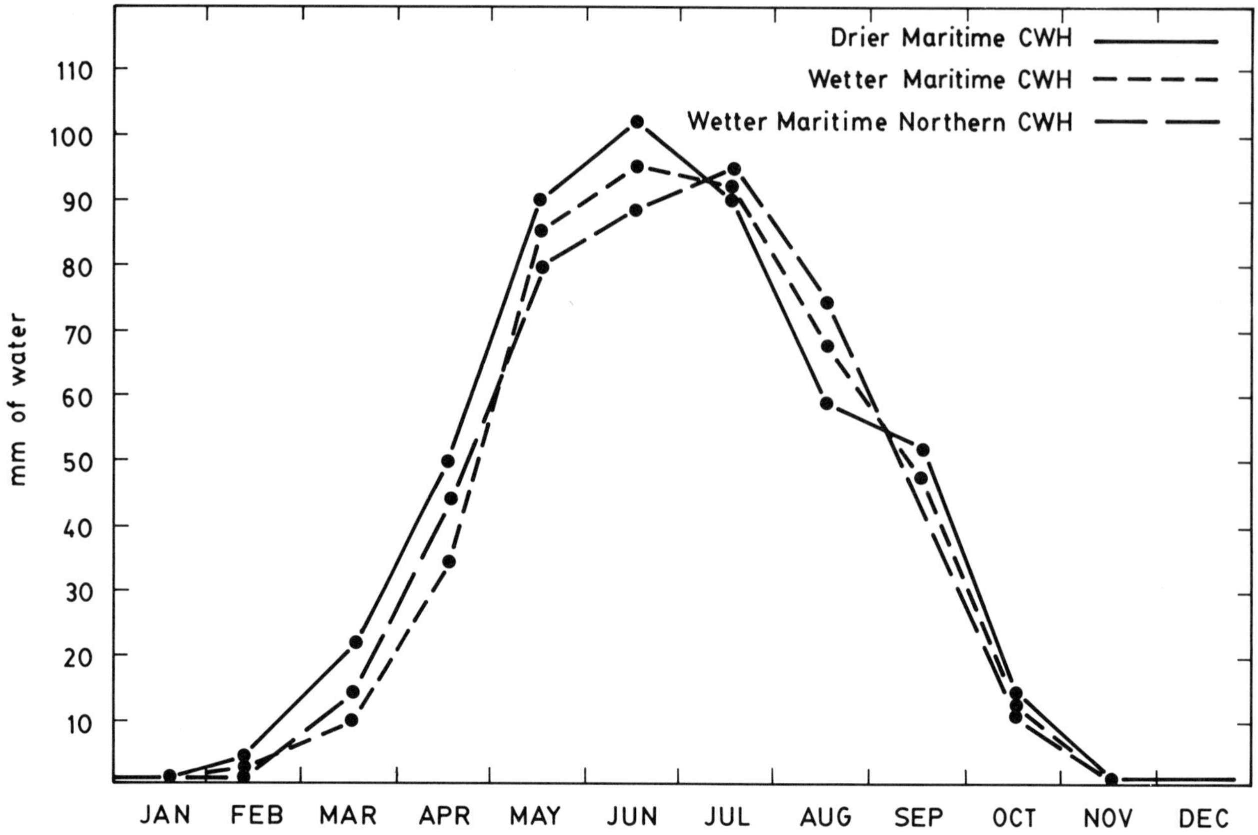

Figure 26. Calculated actual evapotranspiration for the maritime CWH subzones.

discriminate the observations (Table 12). The analysis showed that there is a close relationship between the East Vancouver Island Drier Maritime CWH Variant and the Wetter

Table 12. Means and Standard Deviations of Differentiating Climatic Characteristics for East Vancouver Island and Pacific Ranges Drier Maritime CWH Variants

	Biogeoclimatic Variants	
	East Vancouver Island Drier Maritime CWH	Pacific Ranges Drier Maritime CWH
Number of climatic data sets	6	14
Discriminated Parameters		
Actual evapotranspiration: July (mm)	47 (8.1)	108 (15.3)
Actual evapotranspiration: March (mm)	21 (0.9)	23 (0.7)

Maritime CDF Subzone. This agrees with observations made on the ground. A lack of distinctly different climates, the presence of rich parent materials, and extensive disturbances of the forest cover on eastern Vancouver Island makes the classification and mapping of low elevation ecosystems difficult. In this respect, the discriminant analysis was used to substantiate the classification of a few doubtful areas.

Comparison of actual evapotranspiration values for the variants further supports the above discussion (Figure 27). The curve for the East Vancouver Island Drier Maritime CWH Variant is similar to that for the CDF Zone; for the Pacific Ranges Drier Maritime CWH Variant it resembles the curve for the CWH Zone (Figure 2). The difference is due to a substantial 2.2 month water deficit in the East Vancouver Island Drier Maritime CWH Variant because of low precipitation during the growing season. According to actual evapotranspiration, the Pacific Ranges Drier Maritime CWH Variant supports more intensive plant growth (AET=505 mm as opposed to 432 mm), although the East Vancouver Island Drier Maritime CWH Variant has the potential for higher productivity (PET = 567 mm as opposed to 551 mm) (Appendix IIG).

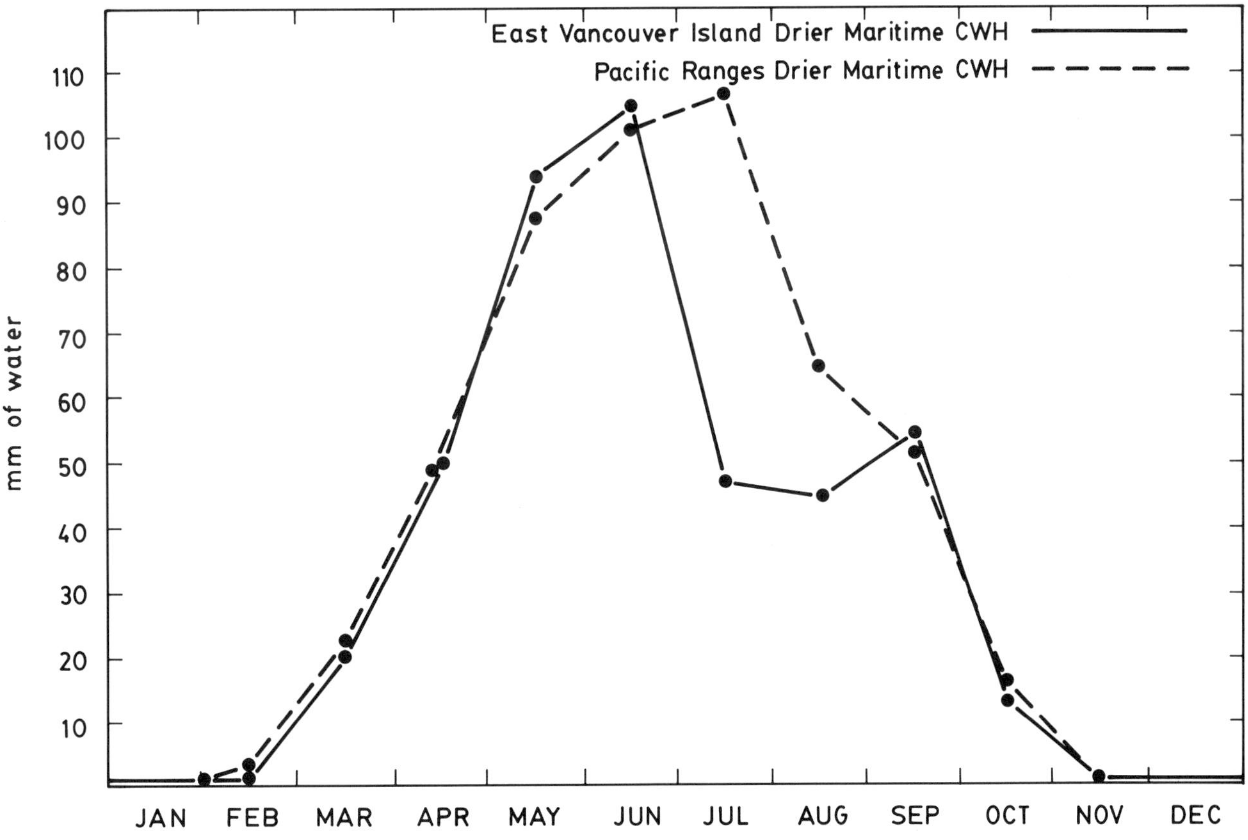

Figure 27. Calculated actual evapotranspiration curves for the East Vancouver Island and Pacific Ranges Drier Maritime CWH variants.

West Vancouver Island Drier Maritime CWH Variant

This variant occurs in the lower parts of major valleys on the western side of Vancouver Island inland from the coast. The climate of these areas, thought to be transitional towards that of the Wetter Maritime CWH Subzone, is to a large degree strongly related to local physiography, *i.e.* high mountain sides oriented along valleys create a rain shadow within the generally wet west coast climate. Thus, this drier and warmer climate has some of the characteristics ascribed to local climates.

On the Nootka-Nanaimo map sheet, the West Vancouver Island Drier Maritime CWH Variant only occurs in the Gold River area. A small area along the road from Gold River to Tahsis between Tlupana River and Conuma River is likely under the influence of a local climate and, consequently, is in the Wetter Maritime CWH Subzone.

Except for very dry sites, western hemlock is found growing with a high species significance and good vigor, especially after windthrow of stands of old growth trees (Figure 28). After such a disturbance on zonal ecosystems, regeneration of Douglas-fir does not occur and the floristic composition of the stand is similar to that of the Wetter Maritime CWH Subzone. Zonal ecosystems are generalized at the level of biogeocoenotic type as: Stokesiella - Rhytidiadelphus - Douglas-fir - Western Hemlock on Humo-Ferric Podzol with friable mor humus. Stands with a larger component of Douglas-fir do exist in the variant. These were established after forest fires and are developing into western hemlock stands through relatively expeditious secondary succession (Figure 29). A few mature stands with a scattered component of grand fir occur on the lower slopes along Heber River and Gold River up to Muchalat Inlet (Figure 30). These stands illustrate that the proposed Douglas-fir, grand fir, and western redcedar mixtures on moist and rich sites are a viable option in the Drier Maritime CWH Subzone. It is interesting to note the occurrence of Pacific madrone and bristly manzanita on very xeric edatopes on rock outcrops facing southwest along Muchalat Inlet. These floristic elements reflect edaphic and microclimatic conditions rather than the existence of a regional climate characteristic of the Wetter Maritime CDF Zone.

East Vancouver Island Drier Maritime CWH Variant

The climate of this variant is akin to that in the Wetter Maritime CDF Subzone. Thus, it marks drier variations of the Drier Maritime CWH Subzone. Ecosystems of this variant occur at middle elevations on the leeward slopes of Vancouver Island mountains. Further away from the east

Figure 28. A vigorous stand of nearly mature western hemlock established following blowdown. High stand density is responsible for great diameter variation and poorly developed crowns. A minor component of Douglas-fir and western redcedar could improve yield [4/C, Stokesiella (oregana) - Rhytidiadelphus (loreus) - PM - TH; Head Bay, Tahsis].

coast of the Island, it occurs at lower elevations (Buttle Lake, Alberni Basin, and Cowichan Lake) and may extend as high at 700 m. Many ecosystems on the upper and middle parts of steep slopes of south and southwest exposures strongly resemble those of the Wetter Maritime CDF Subzone. However, these ecosystems are clearly related to a local climate and are accommodated at the level of the biogeoclimatic phase.

There are very few remaining mature stands available for study. Nevertheless, they are sufficient to provide evidence that western hemlock regenerates with a high species significance under the canopy of forest stands on zonal sites and has the potential to become the climatic climax species (Figures 31, 32, 33, and 34). On the other hand, the vigor of western hemlock on these sites is fair to poor. When mature, it usually exhibits a high level of butt rot and has a dead top. This is due to a combination of relatively dry climate and rich parent materials. Under these same conditions Douglas-fir exhibits

Figure 29. A mature stand of Douglas-fir (dominant) and western hemlock (codominant and suppressed). A greater component of Douglas-fir with an admixture of western redcedar in the understory could improve yield [3/C, Hylocomium (splendens) - Mahonia (nervosa) - PM - TH; Gold River].

good growth.

The biogeocoenotic type, generalizing zonal ecosystems is: Stokesiella - Hylocomium - Douglas-fir - Western Hemlock on incipient Humo-Ferric Podzol with friable mor or moder humus. The scattered occurrence of a few species more common to biogeoclimatic units in the interior of the province (*e.g. Apocynum androsaemifolium, Ceanothus velutinus, Ceanothus sanguineus, Mahonia aquifolium, Paxistima myrsinites,* and *Pleurozium schreberi*), and a high species significance of several species listed in the characteristic combination of species for the Wetter Maritime CDF Subzone (*Amelanchier alnifolia, Gaultheria shallon, Rosa gymnocarpa, Boschniakia hookeri, Chimaphila umbellata,* and *Trachybryum megaptilum*) are characteristic floristic features of the variant.

Figure 30. Part of a vigorous stand of grand fir, Douglas-fir, western redcedar, and western hemlock with advance regeneration of western hemlock, grand fir, and Douglas-fir in shrub layers. However, grand fir is a rare element in forest stands on moist and rich sites in the Drier Maritime CWH Subzone [5/D, Tiarella (trifoliata & laciniata) - Polystichum (munitum) - PM - AG & TP; Gold River].

Douglas-fir maintains some shade tolerance in this variant since it grows in the lower tree and upper shrub layers of stands of second growth trees. It regenerates well on exposed mineral surfaces, but for planting shaded microsites are preferred. The occurrence of many vigorous western white pine trees on richer soils illustrates that it has greater potential as a minor component in stands than is now realized. Pacific madrone is found on very dry, south-facing sites of rock outcrops along Great Central Lake, Sproat Lake, Cowichan Lake, and elsewhere. This is in response to edaphic conditions and a local rather than a regional climate.

High mean radiation and low precipitation during the growing season are responsible for a two month water deficit of 133 mm. As a result, only about 76% of the potential heat is available for plant growth.

Figure 31. A dense stand of mature western hemlock and Douglas-fir in the upper limits of the Drier Maritime CWH Subzone. Western hemlock is a less suitable (vigorous) species as compared to Douglas-fir [4/C, Hylocomium (splendens) - Rhytidiopsis (robusta) - (PM) - TH; Eagle Heights, Shawnigan Lake].

Figure 32. A well-developed blueberry understory in a stand of old growth western hemlock, Douglas-fir, and western redcedar in the transition between the Drier and Wetter Maritime CWH subzones on eastern Vancouver Island. Douglas-fir with a mixture of western redcedar would represent a more suitable stand composition [4/C, Rhytidiadelphus (loreus) - Vaccinium (alaskaense & parvifolium) - (PM) - TH; Ash Lake, Port Alberni].

Figure 33. Abundant and vigorous natural regeneration of western hemlock and western redcedar in the proximity of stands of old growth Douglas-fir indicates that these ecosystems are members of the Drier Maritime CWH Subzone [3/C, Stokesiella (oregana) - Mahonia (nervosa) - PM - TH; Upper Campbell Lake].

Pacific Ranges Drier Maritime CWH Variant

This variant is in the part of the Drier Maritime CWH Subzone that lies along the coastal mainland. The Nootka-Nanaimo map sheet shows a very small area of this variant in the proximity of Powell River and Pender Harbour. In contrast to the East Vancouver Island Drier Maritime CWH Variant, water deficit in August here is small (45 mm). Consequently, actual evaporation and thus, the potential for forest production, is high (505 mm). This value of actual evapotranspiration is only surpassed in the Estevan Submontane Wetter Maritime CWH Variant.

Zonal ecosystems, generalized at the level of biogeocoenotic type are: Hylocomium - Rhytidiadelphus - Douglas-fir - Western Hemlock on Humo-Ferric Podzol with friable mor humus. This variant could be considered as modal for the Drier Maritime CWH Subzone since it could not be provided with a well-defined differentiating combination of species (*i.e.* it has floristic components of both of the other variants).

Figure 34. A dense young stand of immature Douglas-fir, western hemlock (good vigor), and western redcedar exhibiting strong competition in height growth. Juvenile spacing will be required to maintain good growth of the preferred major component of Douglas-fir [4/C, Stokesiella (oregana) & Hylocomium (splendens) - PM - TH; Mohun Lake, Campbell River].

Wetter Maritime CWH Subzone

The Wetter Maritime CWH Subzone is subdivided into seven variants. The submontane and montane variants were further stratified according to West Vancouver Island, East Vancouver Island, and Pacific Ranges physiographic regions to obtain finer climatic differentiation. The Pacific Ranges Wetter Maritime CWH Variant could not be included in the climatic analysis due to the lack of representative data. Finally, the Estevan Submontane Wetter Maritime CWH Variant was added. It is nearly analogous to the Fog Western Hemlock/Sitka Spruce Subzone recognized by Packee (1974, 1976).

In the submontane variants, most of the precipitation falls in the form of rain; and the length of the growing season is not different from that in the Drier Maritime CWH Subzone. In montane variants, winter precipitation is in the form of snow. As a result of its long duration, the growing season is shorter. The bountary between submontane and montane variants is between the elevations of 600 and 700 m.

Discriminant analysis verified the existence of groups within the climatic data that correspond to the variants described above (91.5% of the individual observations were correctly classified). Twenty climatic variables were necessary to discriminate the observations (Table 13). The plotting of canonical variables shows the proximity of some groups to one another as well as a low degree of dispersion of individual observations around the means for the groups. This indicates that affinities exist between the Estevan and West Vancouver Island Submontane Wetter Maritime CWH variants, and between West Vancouver Island and East Vancouver Island Montane Wetter Maritime CWH variants (Figure 35). In contrast, the boundary between submontane and montane variants is more abrupt.

As mentioned earlier, the Wetter Maritime CWH Subzone is characterized by the presence of amabilis fir and yellow-cedar. Both grow best here (particularly on moist and nutrient rich soils) yet they are nearly absent in the drier subzone. The distribution of Douglas-fir is considerably reduced, being related to special habitats or fire history. An abundant supply of moisture promotes establishment and excellent growth of western hemlock over a wide variety of sites. An abundance of mosses (*Rhytidiadelphus loreus, Rhytidiopsis robusta, Plagiothecium undulatum*), the presence of ericaceous shrubs (*Menziesia ferruginea, Vaccinium alaskaense, Vaccinium ovalifolium*), and few herbs (*Clintonia uniflora, Rubus pedatus*) distinguish the understory vegetation of zonal ecosystems.

The course of actual evapotranspiration shows the water deficit in both East Vancouver Island Wetter Maritime CWH variants (Figure 36). This implies that the rain shadow effect occurs over all of the slopes of eastern Vancouver Island. The montane variants have lower actual evapotranspiration rates due to the shorter growing season which commences approximately in the second half of April. No variants of the Wetter Maritime CWH Subzone on west Vancouver Island experience a water deficit. For this reason, potential evapotranspiration is equal to

Table 13. Means and Standard Deviations of Differentiating Climatic Characteristics for the Variants of the Wetter Maritime CWH Subzone

	Biogeoclimatic Variants				
	Estevan Submontane Wetter Maritime CWH	West Vancouver Island Submontane Wetter Maritime CWH	West Vancouver Island Montane Wetter Maritime CWH	East Vancouver Island Montane Wetter Maritime CWH	East Vancouver Island. Submontane Wetter Maritime CWH
Number of climatic data sets	9	11	11	12	4
Discriminated Parameters					
Actual evapotranspiration: March (mm)	25 (1.2)	23 (1.2)	0 (0.0)	0 (0.0)	0 (0.0)
Mean precipitation: April-September (mm)	774 (48.8)	876 (79.0)	967 (135.8)	376 (120.6)	373 (108.2)
Number of months with mean temperature below 0°C	0.0 (0.0)	0.1 (0.30)	0.4 (0.92)	2.7 (0.78)	1.2 (0.50)
Index of continentality	1 (0.2)	2 (4.2)	2 (3.7)	15 (3.1)	15 (2.4)
Mean precipitation of the wettest month (mm)	445 (20.5)	541 (19.3)	597 (55.1)	324 (121.6)	326 (113.6)
Mean precipitation of the driest month (mm)	83 (6.1)	91 (16.9)	102 (20.3)	36 (10.2)	38 (5.3)
Accumulated degree days over 5.6°C	1390 (145.5)	920 (242.0)	502 (118.4)	879 (88.2)	1062 (129.2)
Mean temperature of the warmest month (°C)	14.2 (0.72)	12.4 (1.54)	10.2 (0.98)	13.6 (0.63)	14.8 (0.80)
Mean radiation during growing season (Ly)	40300 (310)	39400 (940)	39600 (1570)	41400 (1740)	41500 (1190)
Water surplus (mm)	2485 (73.4)	3072 (138.1)	3516 (421.3)	1512 (702.2)	1566 (671.9)
Actual evapotranspiration: July (mm)	110 (1.5)	105 (2.0)	103 (5.1)	63 (33.7)	56 (27.5)
Actual evapotranspiration: September (mm)	50 (1.4)	47 (1.1)	45 (1.4)	51 (4.2)	48 (8.7)
Actual evapotranspiration: April (mm)	52 (1.6)	48 (2.0)	23 (0.9)	24 (1.8)	18 (11.9)

Table 13 (Continued)

Actual evapotranspiration: May (mm)	88 (1.8)	83 (2.1)	80 (3.3)	86 (4.9)	90 (2.4)
Number of months with water deficit	0.0 (0.0)	0.0 (0.0)	0.2 (0.60)	2.3 (0.65)	2.5 (0.58)
Maximum snow depth (cm)	0 (0.0)	38 (32.4)	139 (38.3)	130 (71.2)	99 (48.3)
Mean annual temperature (°C)	9.2 (0.43)	7.1 (0.87)	5.0 (0.48)	5.4 (0.40)	6.4 (0.49)
Actual evapotranspiration (mm)	531 (10.7)	502 (8.1)	433 (8.1)	381 (43.1)	388 (54.0)
Actual evapotranspiration: February (mm)	4 (0.8)	4 (0.7)	0 (0.0)	0 (0.0)	0 (0.0)
Number of months with mean temperature over 10°C	5.1 (0.93)	3.8 (0.87)	0.6 (1.03)	3.7 (0.65)	4.0 (0.0)

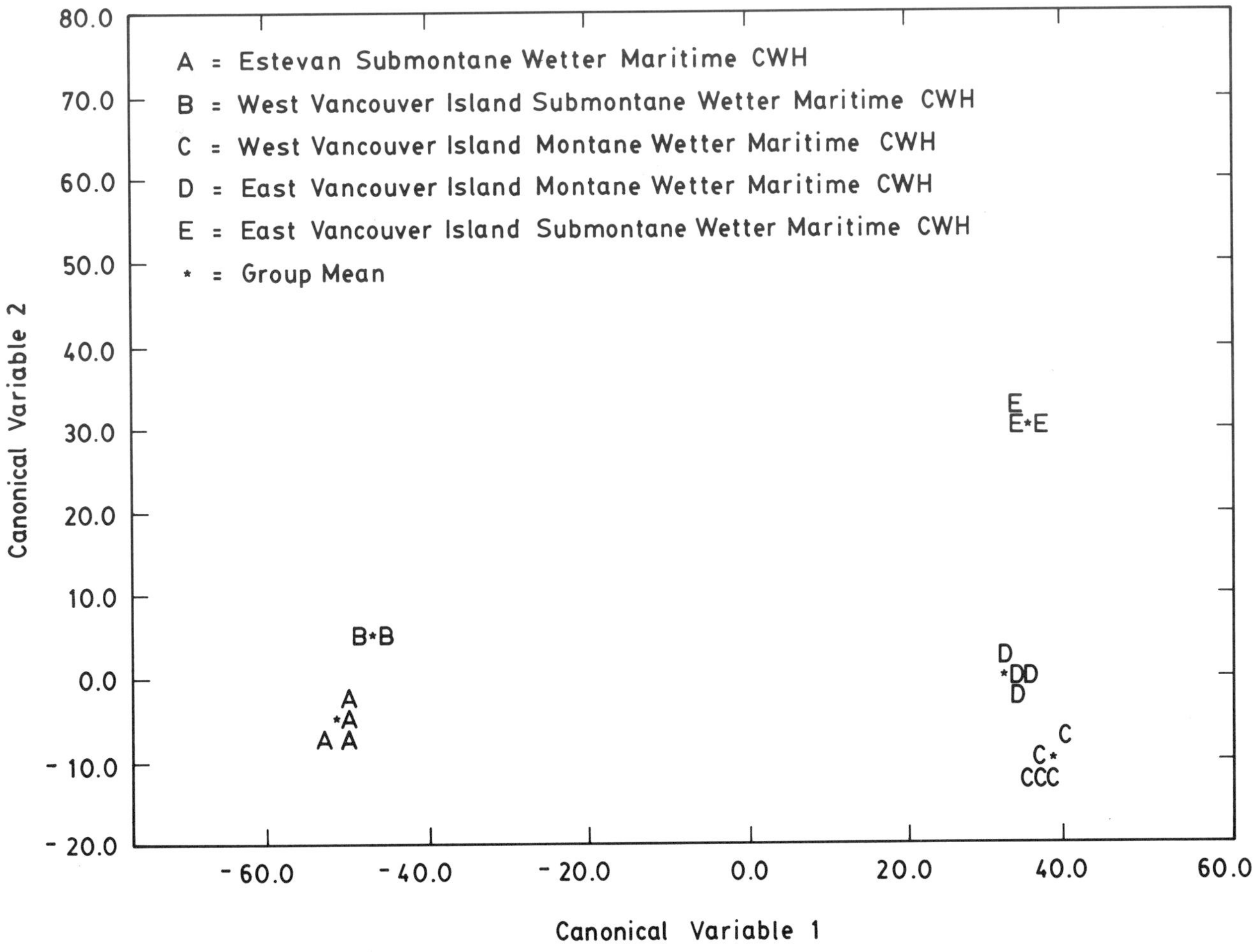

Figure 35. The plot of canonical variables for individual climatic observations which characterize the Wetter Maritime CWH variants.

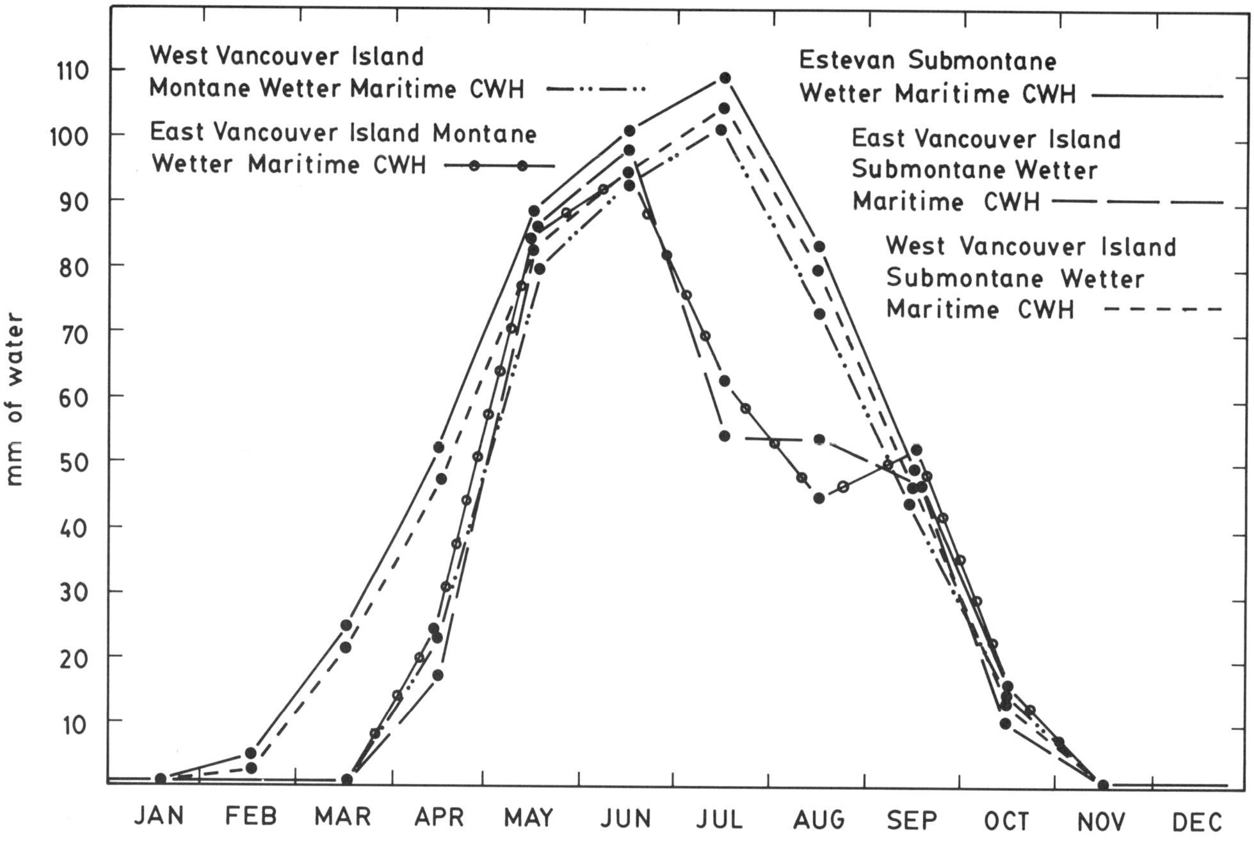

Figure 36. Calculated actual evapotranspiration curves for the variants of the Wetter Maritime CWH Subzone.

actual evapotranspiration in this area. According to the actual evapotranspiration figure of 531 mm, the most intensive plant growth among all of the maritime biogeoclimatic units discussed in this report is in the Estevan Submontane Wetter Maritime CWH Variant.

Estevan Submontane Wetter Maritime CWH Variant

This variant is exclusively limited to a relatively narrow, occasionally discontinuous, belt of lowlands along the outer west coast of Vancouver Island. The width of this belt is approximately 3 to 5 km, extending from River Jordan to Brooks Peninsula. The upper elevation limit may extend to 150 m.

A broad, poorly defined boundary with the neighbouring West Vancouver Island Submontane Wetter Maritime CWH Variant has not definitely been determined due to the lack of consistent differentiating characteristics in the tree strata. The presence of Sitka spruce cannot be used since forest stands where it is a major species are confined to the very narrow band along the coast line which is under the influence of ocean spray (Figure 37) or to alluvial flood-plain deposits of major creeks or rivers (Figure 38). Sitka spruce grows elsewhere, but its abundance is low. Sitka spruce has a similar distribution in the West Vancouver Island Submontane Wetter Maritime CWH Variant. At certain locations the Estevan Submontane Wetter Maritime CWH Variant covers more area than the Fog Western Hemlock / Sitka Spruce Subzone, shown on the biogeoclimatic map of Vancouver Island (MacMillan Bloedel 1974). The ecosystems of the west coast are presently under detailed study by the British Columbia Forest Products Company and the Research Branch of the Forest Service.

Figure 37. An excellent and dense stand of mature Sitka spruce and western hemlock established following windthrow, with prolific advance regeneration of western hemlock in shrub layers (Sitka spruce: age 140 years, mean height 46 m, mean dbh 80 cm; western hemlock: age 132 years, mean height 41 m, mean dbh 76 cm) [5/C, Polystichum (munitum) - Blechnum (spicant) - AA & PS - TH; Esperanza Inlet].

Figure 38. A vigorous Sitka spruce plantation (the recent annual height increment has been approximately 1 m), marginally affected by spruce weevil [5/D-E, Tiarella (laciniata & trifoliata) - Polystichum (munitum) - AA & PS - TP; Kauwinch River, Kyoquot Sound].

The biogeocoenotic type, generalizing zonal ecosystems is: Salal - Red Huckleberry & Alaskan Blueberry - Sitka Spruce & Amabilis Fir - Western Hemlock on Ferro-Humic Podzol with friable mor humus. The characteristic soil features are high accumulation of complexes of iron, aluminum, and amorphous organic matter, indicating an intensive leaching environment and temporary or permanent seepage.

Virtual absence of forest fires has resulted in the development of nearly climatic climax ecosystems. They are characterized by stands of decadent old growth trees including very large, old western redcedar and very dense, tall shrub layers (*Gaultheria shallon, Vaccinium alaskaense, Vaccinium parvifolium, Vaccinium ovalifolium*, and *Vaccinium ovatum*) (Figure 39). Site indices, derived from measurements in these stands, do not represent the true production potential of these ecosystems. The ground surface in old growth stands may show considerable relief which is occasionally over 5 m. This is a reflection of a history of perpetual windthrow of individual trees resulting in great local accumulations of organic matter (mainly wood) in various stages of decomposition. In contrast, areas where extensive windthrow has occurred recently support excellent stands of Sitka spruce and western hemlock or western hemlock and amabilis fir with an admixture of Sitka spruce (Figure 39). With increasing age, the production in these ecosystems deteriorates (Figure 40).

When there is a large water surplus and fine humus in the forest floor, the flat areas and lower parts of gentle slopes are commonly associated with a variety of subhydric sites on which western redcedar, yellow-cedar, and lodgepole pine are the major species. Douglas-fir is a relatively sporadic species, confined to drier and nutrient-rich sites of the upper slopes. For reforestation, Sitka spruce and western hemlock are recommended as the major species with amabilis fir and western redcedar as minor species. Under most circumstances, regeneration by planting is necessary to obtain the desirable tree species composition. Following harvesting of stands of old growth trees where shrub layers are very dense and tall, broadcast burning may be beneficial for expedient regeneration.

West Vancouver Island Submontane Wetter Maritime CWH Variant

Large areas of mountain slopes on western Vancouver Island, between the elevations of 150 and 600 m, are known to receive in excess of 3500 mm of precipitation, making them by far the wettest parts of Canada. Such perhumid climate is characteristic of the West Vancouver Island Submontane Wetter Maritime CWH Variant. It streches across the windward

Figure 39. A stand of old growth western hemlock, western redcedar, and amabilis fir with abundant shrubs and decaying wood. If this were a stand of western hemlock with an admixture of Sitka spruce or western redcedar, it would be more productive [4/C, Gaultheria (shallon) - Vaccinium (parvifolium & alaskaense) - PS & AA - TH; Maggie Lake, Ucluelet].

Figure 40. A stand of old growth amabilis fir and western hemlock featuring a thick layer of decaying wood on and in the forest floor. A stand of Sitka spruce with an admixture of amabilis fir or western hemlock would be more productive [5/C, Polystichum (munitum) - Blechnum (spicant) - AA & PS - TH; Hesquiat Harbour, Estevan Point].

side of Vancouver Island from River Jordan to Quatsino Sound. Westwards it may border on the Estevan Submontane Wetter Maritime CWH Variant or reach the Pacific Ocean.

The biogeocoenotic type, generalizing zonal ecosystems is: Rhytidiadelphus - Red Huckleberry & Alaskan Blueberry - Amabilis Fir - Western Hemlock on Ferro-Humic Podzol with friable mor humus. Under the influence of a large water surplus and lack of heat, mor formation, gleization, accumulation of complexes of organic matter, deposition of iron and aluminum in the lower horizons, and leaching are intensified in the soils. Many of the soils are subjected to the movement of at least temporary seepage through the lower humus horizon or above the restricting layer in the lower parts of the solum.

The infrequent occurrence of forest fires is reflected by the presence of climax ecosystems supporting partly open and very uneven-aged stands. Shrub layers in the old growth stands are very tall and dense; they include *Gaultheria shallon, Menziesia ferruginea, Vaccinium alaskaense, Vaccinium parvifolium,* and *Vaccinium ovalifolium* (Figure 41). Local accumulations of great quantities of slowly decomposing coniferous wood further acidify the soils, thus decreasing nutrient availability. Wind is the most common cause of ecosystem disturbance. In particular, strong southerly, or westerly gale winds are responsible for considerable windthrow, especially on the upper parts of slopes and on ridge tops. This windthrow results in the development of even-aged western hemlock and amabilis fir stands (Figures 42 and 43). The environment in this variant is optimum for the most intensive growth of these two species. Douglas-fir stands are scarce, although a few localities on the west coast have supported or still support stands with a large component of this species (*e.g.* Union Island). These stands are confined mainly to drier and nutrient-rich sites with shallow, sandy -skeletal soils on steep, south-to-southwest facing slopes. Inland from the coast, Douglas-fir generally grows frequently (Figure 44).

West Vancouver Island Montane Wetter Maritime CWH Variant

Higher elevations of the windward slopes of Vancouver Island are under the influence of a wetter and colder mesothermal climate with cool and short summers. The growing season is shorter due to the six month duration of snow cover. The maximum snow depth is estimated to be 139 cm (Appendix IIH). Ecosystems of inland areas show some floristic affinities to the East Vancouver Island Montane Wetter Maritime CWH Variant.

Figure 41. A dense stand of old growth western redcedar, amabilis fir and western hemlock. Western hemlock with an admixture of amabilis fir and western redcedar appears to be the most suitable stand composition [4/C, Rhytidiadelphus (loreus) - Vaccinium (parvifolium & alaskaense) - AA - TH; Loop Creek, Gordon River].

The biogeocoenotic type, generalizing zonal ecosystems is: Rhytidiadelphus - Oval-leaved & Alaskan Blueberry - Amabilis Fir - Western Hemlock on Ferro-Humic Podzol with compacted mycelial mor humus (Figures 45 and 46). Many of the forests have a structure like that seen in even-aged stands and have abundant advance natural regeneration in the shrub layers. On drier sites in the upper limits of the variant or toward the eastern slopes of the Island, *Rhytidiopsis robusta, Gaultheria shallon,* and *Vaccinium parvifolium* are more common. The presence of yellow-cedar and a greater dominance of amabilis fir in the stand composition is the distinct floristic feature of the variant. Western redcedar is also present, decreasing in species significance and vigor as altitude increases. The vigor of western hemlock remains excellent although growth is less vigorous than in the neighbouring submontane variant. Where the environment is optimum for growth of western hemlock, its roots appear to exploit the whole soil solum (Figure 47). Amabilis fir, when favoured by a high nutrient supply, grows with

Figure 42. A stand of vigorous, nearly mature western hemlock, established following windthrow. High density is responsible for great DBH variation and poorly developed crowns; thus potential yield is reduced (4/C, Rhytidiadelphus (loreus) - Vaccinium (parvifolium & alaskaense) - AA - TH; Tofino Inlet).

Figure 43. A dense stand of vigorous, mature amabilis fir and western hemlock, established following windthrow. Amabilis fir with an admixture of western redcedar and Douglas-fir could further increase stand productivity [6/D, Tiarella (trifoliata) - Streptopus (roseus) - AA - TP; Drum Lake, Gold River].

excellent vigor in pure stands even on subxeric and mesic ecosystems (Figure 48). Scattered Douglas-fir is found on xeric edatopes of rock outcrops.

East Vancouver Island Submontane Wetter Maritime CWH Variant

This variant occupies a substantially smaller area than the West Vancouver Island Submontane Wetter Maritime CWH Variant does. At the lower limit, it borders on the Drier Maritime CWH Subzone, and at the upper limit, on the East Vancouver Island Montane Wetter Maritime CWH Variant. It has a discontinuous distribution being confined to areas affected by moderate but relatively long-lasting snow cover. These occur in the upper portions of

Figure 44. A stand of old, yet vigorous, western hemlock and Douglas-fir with a well-developed herb layer in the transition between the Wetter and Drier Maritime CWH subzones. Douglas-fir with an admixture of western redcedar and amabilis fir or grand fir could result in an even more productive stand [4/D, Polystichum (munitum) - Mahonia (nervosa) - (AA) & PM - TP; between Nitinat and Cowichan lakes].

narrow valleys near the highest parts of the mountains of Vancouver Island. A few smaller areas are found along the east slopes of Beaufort Ridge and at Forbidden Plateau. The climate is designated as drier [the annual mean total precipitation is 1904 mm which is less than the 2800 mm used previously to differentiate the wetter from the drier CWH subzones (Krajina 1969)] and colder mesothermal with a short, cool summer. Floristically and environmentally, the variant has considerable affinity with the Drier Maritime CWH Subzone. This is exemplified by a low occurrence of many species common on western Vancouver Island and a water deficit during the growing season.

Figure 45. A dense stand of old growth western hemlock and amabilis fir with abundant advance regeneration. Western hemlock has excellent vigor (4/C, Rhytidiadelphus (loreus) - Vaccinium (ovalifolium & alaskaense) - AA - TH; Poet Heights, Alberni Inlet).

Figure 46. A stand of old growth amabilis fir, western hemlock, and western redcedar with an advanced stage of reproduction in openings or under the canopy [3/C, Rhytidiadelphus (loreus) - Vaccinium (ovalifolium & alaskaense) - AA - TH; Espinosa Inlet].

At the level of biogeocoenotic type, undisturbed zonal ecosystems are generalized as: Rhytidiadelphus - Red Huckleberry & Alaskan Blueberry - (Amabilis Fir) - Western Hemlock on Humo-Ferric (rarely Ferro-Humic) Podzol with mycelial mor humus. Nearly all mature forest stands outside Strathcona Provincial Park were logged and many of those within the Park were burned. As a result, Douglas-fir and the understory vegetation associated with it have come to dominate the flora of disturbed stands within this variant. Douglas-fir grows well here in spite of the snow cover and a shorter growing season. In contrast, due largely to an inadequate moisture supply during the growing season, the growth of both western hemlock and amabilis fir is slow. The re-establishment of these two species could be exceedingly slow where advance regeneration and the humus are destroyed by fire. Where amabilis fir is present in disturbed old growth stands, it has low species significance and is mostly in the lower tree and shrub layers.

Figure 47. A road cut at an elevation of 750 m featuring a very thin layer of soil solum (about 30 cm), developed from loose till overlying compacted till [4/C, Rhytidiadelphus (loreus) - Vaccinium (ovalifolium & alaskaense) - AA - TH; Carmanah Mountain, Nitinat Lake].

Figure 48. A dense stand of old growth amabilis fir featuring excellent vigor due to an abundant nutrient supply, provided by soils derived from limestone [4/E, Tiarella (trifoliata) - Streptopus (roseus) - TP & AA; Sucwoa River, Tlupana Inlet].

A sporadic occurrence of stands of young trees with a floristic composition approximating that of the forests of old growth trees was helpful for determining the extent of the variant (Figures 49, 50, 51, and 52). In addition, the presence of *Gaultheria shallon*, *Mahonia nervosa*, *Rosa gymnocarpa*, *Achlys triphylla*, *Chimaphila menziesii*, *Chimaphila umbellata*, and *Trachybryum megaptilum* was beneficial for differentiating it from the East Vancouver Island Montane Wetter Maritime CWH Variant.

Figure 49. An open stand of immature amabilis fir and western hemlock originating from advance regeneration. Both amabilis fir and western hemlock have poor vigor. Douglas-fir with an admixture of western redcedar and white pine would be a more productive option [4/C, Rhytidiadelphus (loreus) - Vaccinium (parvifolium & alaskaense) - (AA) - TH; Mount Washington, Courtenay].

Figure 50. A fragment of a stand of immature amabilis fir at an elevation of 230 m. Amabilis fir exhibits poor to fair vigor [2/C, Rhytidiadelphus (loreus) - Vaccinium (parvifolium & ovalifolium) - AA & PM - TH; east of Paterson Lake, Campbell River].

Figure 51. A dense stand of immature amabilis fir with fair vigor due to additional moisture supply as seepage. A substantial Douglas-fir component in the stand could be more productive [5/C, Rhytidiadelphus (loreus) - Blechnum (spicant) - AA - TH; Elsie Lake, Port Alberni].

Figure 52. An open stand of immature western hemlock and amabilis fir of fair to good vigor. A major component of Douglas-fir with an admixture of amabilis fir and western hemlock would be a more productive option [5/C, Rhytidiadelphus (loreus) - Blechnum (spicant) - AA - TH; Dickson Lake, Port Alberni].

East Vancouver Island Montane Wetter Maritime CWH Variant

This variant is found at higher elevations on the leeward slopes of Vancouver Island between 700 m and 1100 m. The climate is very much like that of the East Vancouver Island Submontane Wetter Maritime CWH Variant except the growing season is shorter and the winter more severe. Floristically and environmentally, this variant has some affinities with the Drier Maritime CWH and the Wetter Subcontinental CWH subzones. These are manifested by the absence or scattered occurrence of species commonly identified with the Wetter Maritime CWH Subzone, the presence of a few interior species (*Abies lasiocarpa, Pinus monticola, Gaultheria ovalifolia, Paxistima myrsinites,* and *Rhododendron albiflorum*) and a water deficit during the growing season. The few remaining undisturbed zonal ecosystems are usually located in the upper limits of the variant. At the level of biogeocoenotic type they are classified as: Rhytidiopsis - Oval-leaved & Alaskan Blueberry - Amabilis Fir - Western Hemlock on Humo-Ferric (rarely Ferro-Humic) Podzol with compacted mycelial mor humus. The estimated maximum winter snow depth is between 60 and 360 cm with a mean value of 130 cm. The differences in vegetation characteristics and ecosystem productivity are clearly related to local or altitudinal differences in snow depth and duration of snow cover (Figures 53 and 54).

For the most part, the forests have been significantly affected by fire. This has resulted in suppressed regeneration and vigor of western hemlock and particularly amabilis fir. Thus, stands of mixture trees are in various successional stages, usually with a large component of Douglas-fir and little amabilis fir (Figure 55). They include a few floristic elements common to the lower elevations (*Gaultheria shallon, Mahonia nervosa, Vaccinium parvifolium, Chimaphila menziesii, Chimaphila umbellata,* and *Trachybryum megaptilum*). The presence of yellow-cedar and amabilis fir in the lower tree or brush layers of zonal ecosystems is a reliable identification characteristic being particularly useful for defining the lower limits of the variant. Providing no broadcast burning is carried out, amabilis fir and western hemlock may regenerate with some delay on cut-over areas. However, both species grow with fair to poor vigor (Figure 56). Subalpine fir grows on very dry and nutrient rich sites of basaltic outcrops and cliffs.

Figure 53. A stand of old growth amabilis fir and western hemlock with a little white pine, Douglas-fir, and yellow-cedar at 760 m elevation. Only a minor component of Douglas-fir should be retained due to excessive snow cover (as high as 360 cm) [4/C, Rhytidiopsis (robusta) - Vaccinium (ovalifolium & alaskaense) - AA - TH; Sadie Creek, Nanaimo River].

Figure 54. A dense stand of old growth western hemlock with a minor component of amabilis fir, western redcedar, yellow-cedar, and mountain hemlock in the upper limits (940 m) of the variant [5/C, Rubus (pedatus) - Vaccinium (ovalifolium & alaskaense) - TP & AA - TH; Robertson River].

Pacific Ranges Montane Wetter Maritime CWH Variant

Montane ecosystems of the Wetter Maritime CWH Subzone along the coastal mainland are included in this variant. The Cfc climate is believed to be similar to, or perhaps slightly drier than, that of the West Vancouver Island Montane Wetter Maritime CWH Variant. On the Nootka-Nanaimo map sheet, there is a very small area of the Pacific Ranges Montane Wetter Maritime CWH Variant in the proximity of Powell River and Pender Harbour.

The biogeocoenotic type, generalizing zonal ecosystems is: Rhytidiadelphus - Oval-leaved & Alaskan Blueberry - Amabilis Fir - Western Hemlock on Humo-Ferric or Ferro-Humic Podzols with compacted mycelial mor humus. The approximate normal winter snow depth is estimated to be over 150cm.

Figure 55. A stand of old growth Douglas-fir, western hemlock, yellow-cedar, western redcedar, and amabilis fir at the lower limits (810 m) of the variant. Use of Douglas-fir, with a minor component of western redcedar or yellow-cedar and white pine would result in increased forest productivity. Judging by the height of lichen growth on the trunks of trees, snow cover was estimated to be 140 cm [4/C, Rhytidiopsis (robusta) - Vaccinium (ovalifolium & alaskaense) - AA - TH; Nanaimo lakes].

Figure 56. A stand of immature western hemlock, Douglas-fir, amabilis fir, and yellow-cedar on a northeast slope at 670 m elevation [4/C, Rhytidiopsis (robusta) - Vaccinium (ovalifolium & alaskaense) - AA - TH; Mount Benson, Nanaimo].

APPLICATION

In the application of science and technology to ecosystems, there is a need to simplify the complexity of nature's many details to permit an efficient organization of our knowledge and a clear perception of relationships. This is done in ecosystematic classification where ecosystems are grouped into classes at various levels of generalization (categories). The biogeoclimatic level is most generalized in this classification. Few specific statements can be made about its categories, but general statements or specialized applications are possible.

Biogeoclimatic units are a tool for communicating accumulated observations, evaluations and experience, and for making general predictions. They make it possible to stratify sample plots for inventory and various studies concerning resource management. Biogeoclimatic zones, subzones, and variants provide benchmarks for further comparative ecological studies of ecosystems and their components. For instance, a sequence of ecosystems, dominated by a single factor, *e.g.* by climate, could be sought to determine the relationship between ecosystems and climate. These studies would supply information about such things as the general climatic requirements of plants and their ecological function, and allow comparisons between the soils to elucidate their formation.

The most common use of biogeoclimatic units is as a framework for site (ecosystem)-specific management. Each biogeoclimatic zone, subzone, or variant has its own set of plant orders, biogeocoenotic associations and biogeocoenotic types respectively, distributed in a certain pattern in the landscape. Therefore, the biogeoclimatic units together with the map become an index for this set of plant orders and biogeocoenotic units. Certain interpretations and prescriptions concerning these can be made. For example, biogeoclimatic subzones and the pattern of biogeocoenotic associations within them are used as a framework in guides for tree species selection and prescribed burning (Wyeth and Klinka 1975, Klinka 1977, Utzig and MacDonald 1977). In general, where aspects of forest, range, wildlife, watershed, and recreation management are related to ecosystems, and the relationships have been studied, it is possible to apply biogeoclimatic classification. At the present time, no field of management other than forestry has received much attention in this regard.

The biogeoclimatic level is part of a taxonomic classification used to classify ecosystems without modifications to accommodate the user. For some applications in management, simplification is possible. This is accomplished by grouping taxonomic units into a few interpretive classes named treatment units (Klinka 1976, Resource Planning Division 1976, Krajina, 1977). The way certain biogeocoenotic units respond to management practices necessitates that they be regarded as special treatment units; in other cases, a number of biogeocoenotic units (*e.g.* one association and three types) or even biogeoclimatic units (*e.g.* a subzone and an adjacent variant from another subzone) can be fused into a single treatment unit. The criteria for delineating a treatment unit may vary between fields of management. Thus, a treatment unit for forestry may encompass more or fewer biogeocoenotic units or biogeoclimatic units than a treatment unit for range management. Treatment units for forest management are delineated where the vegetation-environment relationships, productivity of biogeocoenotic units, and optimum combinations of tree species are similar. In the coastal part of the Vancouver Forest Region, eight forested and one non-forested treatment units are distinguished (Figure 57).

Depending on the requirements of the user, either taxonomic (*i.e.* the category of biogeoclimatic zone, subzone, or variant) or treatment units can be employed. For instance, the biogeoclimatic variant is suitable for governing seed collection and transfer of planting stock in a tree improvement program since, as shown in this report, it delineates areas with comparatively little variation in regional climate. The application of treatment units would be more suitable for forest protection (fire weather zones), forest inventory (forest inventory zones), growth and yield studies [local yield tables, productivity zones (Schmidt 1971)], and the intensive forestry program (regional silvicultural regimes and practices, management priorities, and evaluation of stand history). Treatment units will be used in the third approximation of the guide for tree species selection and site preparation in the Vancouver Forest Region. They will allow more efficacious silvicultural interpretations than the taxonomic units.

<table>
<tr><th colspan="3">Biogeoclimatic Categories</th><th></th></tr>
<tr><th>Zone</th><th>Subzone</th><th>Variant</th><th>Treatment Unit Number</th></tr>
<tr><td rowspan="3">Coastal Douglas-fir</td><td>Drier Maritime CDF</td><td>Nanaimo & Georgia Drier Maritime CDF</td><td rowspan="4">1</td></tr>
<tr><td rowspan="2">Wetter Maritime CDF</td><td>Nanaimo & Georgia Wetter Maritime CDF</td></tr>
<tr><td>Fraser Wetter Maritime CDF</td></tr>
<tr><td rowspan="11">Coastal Western Hemlock</td><td rowspan="3">Drier Maritime CWH</td><td>East Vancouver Island Drier Maritime CWH</td></tr>
<tr><td>Pacific Ranges Drier Maritime CWH</td><td rowspan="2">2</td></tr>
<tr><td>West Vancouver Island Drier Maritime CWH</td></tr>
<tr><td rowspan="7">Wetter Maritime CWH</td><td>Estevan Submontane Wetter Maritime CWH</td><td rowspan="2">3</td></tr>
<tr><td>West Vancouver Island Submontane Wetter Maritime CWH</td></tr>
<tr><td>East Vancouver Island Submontane Wetter Maritime CWH</td><td rowspan="2">4</td></tr>
<tr><td>East Vancouver Island Montane Wetter Maritime CWH</td></tr>
<tr><td>Pacific Ranges Submontane Wetter Maritime CWH</td><td rowspan="3">5</td></tr>
<tr><td>West Vancouver Island Montane Wetter Maritime CWH</td></tr>
<tr><td>Pacific Ranges Montane Wetter Maritime CWH</td></tr>
<tr><td>Wetter Maritime Northern CWH</td><td></td><td>6</td></tr>
<tr><td rowspan="4">Mountain Hemlock</td><td rowspan="3">Maritime Forested MH</td><td>East Vancouver Island Maritime Forested MH</td><td>7</td></tr>
<tr><td>West Vancouver Island Maritime Forested MH</td><td rowspan="2">8</td></tr>
<tr><td>Pacific Ranges Maritime Forested MH</td></tr>
<tr><td>Maritime Parkland MH</td><td></td><td rowspan="2">9</td></tr>
<tr><td>Alpine Tundra</td><td>Coastal AT</td><td></td></tr>
</table>

Figure 57. Proposed treatment units for the maritime part of the Vancouver Forest Region.

BIOGEOCLIMATIC MAPPING

The objectives in preparing the map that accompanies this report were to delineate the biogeoclimatic units found on the area covered by the Nootka-Nanaimo map sheet for the categories of zone, subzone, and variant, and to propose a design for a provincial biogeoclimatic map series. The results of the second objective are apparent in the map; those of the first require elaboration.

Prior to any mapping, a synopsis of biogeoclimatic units was prepared. For each category a list of the characteristic combination of species was then made using taxonomic data and other information of Krajina and Spilsbury (1952, 1953), Krajina (1959, 1965, 1969, 1972, 1975a & b, 1976, 1977), McMinn (1957), Mueller-Dombois (1959), Lesko (1961), Orloci (1961, 1965), Archer (1963), Wade (1965), Brooke *et al.* (1970), Kojima (1971), Schmidt (1971), Cordes (1972), Roehmer (1972), Klinka (1976), Lewis (1976), and Packee (1976).

The mapping was done by a confirmation of ground and aerial surveys during the summer and fall of 1977. Boundaries were drawn according to the occurrence of zonal ecosystems and the presence or absence of the plant species (trees, shrubs, herbs, mosses, liverworts, and lichens) of the characteristic combinations, which were prepared beforehand. The map of biogeoclimatic zones (Krajina 1973), the map of biogeoclimatic subzones of Vancouver Island and adjacent mainland (MacMillan Bloedel 1974), and the maps of biogeoclimatic units of Pacific Logging Company (Roehmer and Korelus 1975) were consulted during mapping. Boundaries were finalized following analysis of climatic data and its synthesis with vegetation-environment data obtained for representative zonal ecosystems. Where boundaries were not checked at selected spots on the ground or from the air, they were extrapolated using a combination of elevation and physiography. Due to inaccuracy resulting from this extrapolation, as much as 15% of a mapping unit may belong to an adjacent unit. This is one reason why ground checking by the user and reference to the information that surrounds the map is necessary for correct identification of the biogeoclimatic unit within a mapping unit.

With a few exceptions the locations of variants are delineated as individual mapping units. In the Coastal AT Subzone, no variants were differentiated, so it is a single mapping unit. The Maritime Parkland MH Subzone and the Maritime Forested MH variants are combined into a composite mapping unit because it was not possible to map the parkland within the limits of accuracy desired. The mapping unit was separated into the West Vancouver Island, East Vancouver Island,

and Pacific Ranges geographical areas. The other composite mapping unit includes the Estevan Submontane and the West Vancouver Island Wetter Maritime CWH variants. These variants are related ecologically and geographically, and therefore are difficult to separate without a prohibitive amount of ground checking in areas where accessibility is limited. When a composite mapping unit is encountered in the field and the user wishes to identify biogeoclimatic units, it is only necessary to note a few ground features and refer to the information surrounding the map. For example, the Estevan Submontane Maritime Forested MH Variant on the western coast of Vancouver Island is located where the elevation is roughly 150 m or less, where *Picea sitchensis, Vaccinium ovatum, Usnea longissima,* and less frequently, *Gaultheria shallon, Vaccinium parvifolium, Calamagrostis nutkaensis,* and *Polypodium scouleri* are found together on zonal ecosystems, and where *Pinus contorta* grows in other than zonal ecosystems.

Ecosystems near the boundary of a mapping unit will always have some characteristics of those of the neighboring unit. In mountainous areas of southwestern British Columbia, the transition from one mapping unit to another may occur over a horizontal distance of 50 m if there is a steep altitudinal gradient, while along latitudinal or longitudinal gradients the transition is usually greater than one kilometer. Due to the inaccuracy involved in extrapolation and the nature of the boundaries between mapping units, the scale 1:500 000 (1 mm = 500 m) has been found an acceptable compromise between the intended function of the map and its costs. The definition between mapping units along steep altitudinal gradients would be increased if a larger map scale were used. However, there would be no advantage in the areas where topography is more subdued. On the other hand, a distinct disadvantage would be the great increase in cost With the additional information surrounding the map and that contained in this report it is possible and often necessary, as discussed above, to go to the field to determine the boundary between the biogeoclimatic units. The accuracy thus attained would be greater than that possibl on the basis of a map examination alone, regardless of whether it is at a scale of 1:500 000 or larger.

CARTOGRAPHIC DESIGN CONSIDERATIONS

by

Lou Skoda

The maps of the biogeoclimatic units are intended to serve as aids in planning and in control of organic production in forest management. Because of their innovative approach, they are also expected to function as aids in the general educational process. The third major consideration affecting design envisaged that individual sheets were to form a series that would eventually cover the whole province.

To meet these objectives, individual map sheets were assigned a substantial amount of text, explanations, and illustrations so that each would function as a self contained publication. In addition, the symbolization and, more specifically, the color scheme representing the biogeoclimatic units had to apply to the whole series. Other major design specifications related to scale and to the base map format.

The scale of 1:500 000 was chosen because it appeared to be large enough to allow clear representation of all the expected field data, yet small enough to make the mapping of the province in 23 map sheets, a relatively minor cartographic undertaking. The existing federal government 1:500 000 series, produced by the Department of Energy, Mines, and Resources, conveniently serves as the base map with appropriate modifications. The map sheets in this series will normally occupy an area bounded by 2° of latitude and 4° of longitude as determined by the National Topographic System sheet designations, although this size will be increased for some sheets along the coastline to minimize the number of sheets that would otherwise be required. The biogeoclimatic series will consist of 23 map sheets as shown by the "Key to Sheets" on the map.

Color design

The major thrust of the cartographic design work for the Nootka Nanaimo map sheet focussed on preparation of suitable color sequences to represent the biogeoclimatic units in the whole series.

The principal concepts of assigning color in vegetation mapping have been formulated by H. Gaussen (1960, 1970) and elaborated on in other publications (*e.g.* UNESCO 1973, Kuchler 1967, Mueller-Dombois and Ellenberg 1974, Sochova and Isachenko 1973). The concepts which are relevant to the present project concern assignment of color, *i.e.* hues, to mapping units according to their characteristic climate.

Experimentation with alternative schemes led to a choice of hues listed in Figure 58a and illustrated in a Munsell hue and chroma diagram (Figure 58b). These hues are listed in

a

biogeoclimatic zone		*mean annual temperature* °C	*climate type (Köppen - Trewartha)*	*Munsell colour designation hue/chroma*
Alpine Tundra	(AT)	$-3^{1}/_{2}$	ET	5PB /2
Boreal White and Black Spruce	(BWBS)	0	Dfc	10B /4
Sub-Boreal Spruce	(SBS)	2	Dfc	5B /6
Spruce - Willow - Birch	(SWB)	$2^{1}/_{2}$	Dfc	10BG /7
Engelmann Spruce - Subalpine Fir	(ESSF)	$2^{1}/_{2}$	Dfc	5BG /8
Mountain Hemlock	(MH)	5	Dfc	7.5G /10
Interior Western Hemlock	(IWH)	$5^{1}/_{2}$	Dfb	10GY /10
Interior Douglas - fir	(IDF)	$6^{1}/_{2}$	Dfb	5GY /12
Coastal Western Hemlock	(CWH)	7	Cfb	10Y /9
Coastal Douglas - fir	(CDF)	10	Csb	5Y /12
Ponderosa Pine - Bunchgrass	(PPBG)	8	Bsk	7.5R /12

b

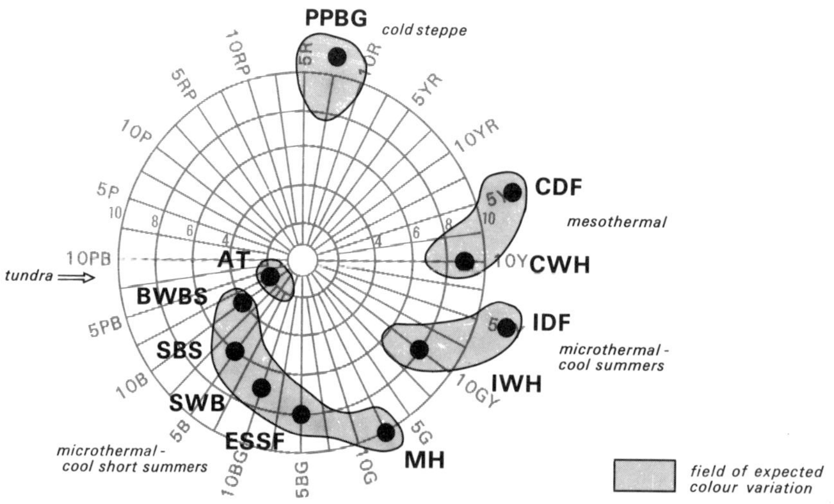

Figure 58a and b. Early color scheme for the biogeoclimatic subzone series and the corresponding definition on the Munsell color diagram.

general agreement with what could be regarded as international convention and their ordering follows a sequence determined by temperature. From a color proof of a pilot sheet, however, it became evident that the color selection was not the most appropriate for the British

Columbia series. The closeness of the hues meant that the representation of subzones and variants, up to 10 or more for each zone, with clearly discernible differences but still recognizable as belonging to a single hue, would be difficult to attain. In addition, the mean annual temperatures used were based on scanty data with inherent unreliability. With the acquisition of new climatic data, it could have become necessary to change the ranking. Further, the choice utilized only a small segment of the possible useful colors, mainly in the blue-green-yellow segment, limiting the potential aesthetic and clarity aspects of the series. Therefore the color scheme was changed to eliminate these shortcomings.

The final color scheme devised for the series is illustrated in Figure 59a and b. The color scheme reflects three climatic groupings - cool, warm and dry, and warm and moist.

a

biogeoclimatic zone		*climate*	*productivity*	*Munsell colour designation hue/chroma*
Alpine Tundra	(AT)	cool	*increasing* ↓	7.5G /1
Spruce - Willow - Birch	(SWB)			7.5G /2
Boreal White & Black Spruce	(BWBS)			7.5B /8
Sub - Boreal Spruce	(SBS)			5PB /9
Mountain Hemlock	(MH)			5P /10
Engelmann Spruce - Subalpine Fir	(ESSF)			2.5RP /11
Ponderosa Pine - Bunchgrass	(PPBG)	dry & warm		2.5R /11
Interior Douglas - fir	(IDF)			5YR /9
Coastal Douglas - fir	(CDF)	moist & warm		5Y /9
Interior Western Hemlock	(IWH)			5GY /10
Coastal Western Hemlock	(CWH)			5G /11

b

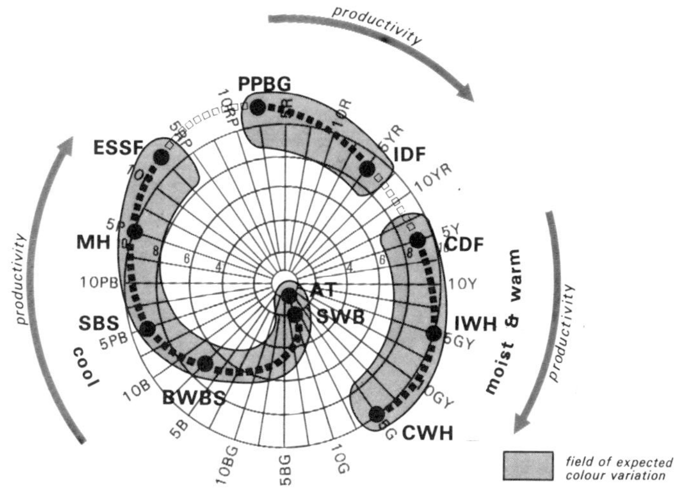

Figure 59a and b. Color schemes for the biogeoclimatic subzone series and the corresponding definition on the Munsell color diagram.

Also the ordering of the zones in the scheme generally reflects productivity. On the Munsell color hue and chroma diagram (Figure 59b) productivity increases in a clockwise direction although certain individuals of the three climatic groups may have similar productivities *i.e.* the productivity ranking is independent in each of the climatic groupings, but the direction of the increase remains the same.

The color samples from the Munsell color solid assigned to each zone represent ideal color sequences which will need to be translated into operational/printing terms. In this translation the hues are expected to deviate from the sample to some extent. The expected magnitude of the deviation is indicated in Figure 59b as a shaded area.

The principal objective of the selected hue sequence is to retain the existing organizing framework of the biogeoclimatic zones throughout the series. In preparing the earlier proofs of the Nootka-Nanaimo sheet, subzones and variants as well as the zones were assigned a combination of chroma, texture, and overprint that attempted to completely predetermine the way any of the mapping units would have to be represented throughout the series. This approach, consistent with past vegetation mapping practices, is logical and clear as an organizing procedure, but may fall short of satisfying the basic design requirements of considering the size and juxtaposition of each map unit for maximum visual effectiveness and clarity. The necessary differentiation between subzones and variants is now achieved through chroma and texture but the process of selection is based on the distribution of mapping units within the sheet boundaries. The procedure in designing subsequent sheets will therefore need to include consideration of previous decisions so that continuity across sheet boundaries can be maintained.

LITERATURE CITED

Archer, A.C. 1963. Some synecological problems in the alpine zone of Garibaldi Park. University of British Columbia, Vancouver. B.C. 129 p. Thesis.

Beil, C.E., R.L. Taylor, and G.A. Guppy. 1976. The biogeoclimatic zones of British Columbia. Davidsonia 7: 45-55.

Black, T.A. [1979]. Evapotranspiration from Douglas-fir stands exposed to soil water deficits. Water Resources Research. In press.

Braun-Blanquet, J. 1928. Pflanzensoziologie. Grundzüge der Vegetationskunde. 1 Auflage. Springer, Berlin. 330 p.

______________. 1932. Plant Sociology. McGraw-Hill Book Co. Inc., New York. 439 p. Translated from the German by H.S. Conard and G.D. Fuller.

Brooke, R.C., E.B. Peterson, and V.J. Krajina. 1970. The Subalpine Mountain Hemlock Zone. Ecology of Western North America 2(2): 148-349.

Cline, M.G. 1949. Basic principles of soil classification. Soil Sci. 67: 81-91.

Cordes, L.D. 1972. An ecological study of the Sitka spruce forest on the west coast of Vancouver Island. University of British Columbia, Vancouver. B.C 452 p. 2 Vol. Dissertation.

Crum, A.H., W.C. Steere, and L.E. Anderson. 1973. A new list of mosses of North America, north of Mexico. The Bryologist 76: 85-130.

Duhem. P. 1974. The aim and structure of physical theory. Atheneum, New York. 344 p. Translated by P.P. Wiener.

Gaussen, H. 1960. L'emploi des couleurs dans la cartografie de la végétation Pp. 140-143 in Méthodes de la cartografie de la végétation. Toulouse Colloquium, Paris, France.

__________. 1970. Vegetation map of the Mediterranean Zone. United Nations Educational, Scientific, and Cultural Organization. Paris, France. 82 p.

Hale, M.E., Jr. and W.L. Culberson. 1970. A second checklist of the lichens of the continental United States and Canada. The Bryologist 63: 137-172.

Halm, J. 1970. Stepwise discriminant analysis. Computing Centre, University of British Columbia, Vancouver, B.C. 14 p. Documentation: UBC BMD07M. Revised 1975.

Holland, S.S. 1976. Landforms of British Columbia, a physiographic outline. British Columbia Department of Mines and Petroleum Resources, Victoria, B.C. Bulletin 48. 138 p.

Klinka, K. 1976. Ecosystem units, their classification, interpretation, and mapping in the University of British Columbia Research Forest. University of British Columbia, Vancouver, B.C. 622 p. Dissertation.

_________. 1977. Guide for the tree species selection and prescribed burning in the Vancouver Forest District. 2d approximation. Ministry of Forests. British Columbia Forest Service, Research Division, Vancouver Forest District, Vancouver, B.C. 42 p.

Kojima, S. 1971. Phytogeocoenosis of the Coastal Western Hemlock Zone in Strathcona Provincial Park, British Columbia, Canada. University of British Columbia. Vancouver, B.C. 321 p. Dissertation.

Krajina, V.J. 1959. Biogeoclimatic zones in British Columbia. University of British Columbia, Vancouver, B.C. Botanical series no. 1. 47 p.

___________. 1960a. Can we find a common platform for the different schools of forest type classification? Silva Fennica 105: 50-55.

___________. 1960b. Ecosystem classification of forests. Silva Fennica 105: 107-110. 123-138.

___________. 1965. Biogeoclimatic zones and classification of British Columbia. Ecology of Western North America 1: 1-17.

___________. 1969. Ecology of forest trees in British Columbia. Ecology of Western North America 2(1): 1-146.

___________. 1972. Ecosystem perspectives in forestry. University of British Columbia, Vancouver, B.C. 31 p. Paper presented as the H.R. MacMillan Lectureship in Forestry.

___________. 1973. Biogeoclimatic zones of British Columbia; Department of Lands, Forests, and Water Resources, Victoria, B.C. British Columbia Ecological Reserves Committee; 1:1 900 800; 94 x 77 cm; colored map.

___________. 1975a. Some observations on the three subalpine biogeoclimatic zones in British Columbia, Yukon, and MacKenzie District. Phytocoenologia 2(3/4): 396-400.

___________. 1975b. Revision of biogeoclimatic zonation of British Columbia. 3 p. Paper located at: Department of Botany, University of British Columbia, Vancouver, B.C.

___________. 1976. Biogeoclimatic zones of British Columbia. 11 p. Paper located at: MacMillan Bloedel Place, VanDusen Botanical Gardens, Vancouver, B.C.

___________. 1977. On the need for an ecosystem approach to forest land management. P. 1-11 in Ecological classification of forest land in Canada and northwestern U.S.A. A symposium by the Forest Ecology Working Group of the Canadian Institute of Forestry and the Centre for Continuing Education, the University of British Columbia; 30 September 1977. Available from: Centre for Continuing Education, University of British Columbia, Vancouver, B.C.

___________. 1978. Vegetation of western North America. 23 p. Paper located at: Department of Botany, University of British Columbia, Vancouver, B.C.

___________. and R.H. Spilsbury. 1952. The ecological classification of the forests of the eastern part of Vancouver Island. Interim report. 51 p. Paper located at: Department of Botany, University of British Columbia, Vancouver, B.C.

_______________________. 1953. Forest associations on the east coast of Vancouver Island. P. 142-145 in Forestry handbook for British Columbia. The Forest Club. University of British Columbia, Vancouver, B.C.

Kuchler, A.W. 1967. Vegetation mapping. Ronald Press Co., New York. 472 p.

Lavkulich, L.M. 1972. Soil classification and mapping. 4 p. Paper P-1/72 located at: Dept. of Soil Science, University of British Columbia, Vancouver, B.C.

Lesko, G.L. 1961. Ecological study of soils in the Coastal Western Hemlock Zone. University of British Columbia, Vancouver, B.C. 114 p. Thesis.

Lewis, T. 1976. The till-derived podzols of Vancouver Island. University of British Columbia, Vancouver, B.C. 158 p. Dissertation.

MacMillan Bloedel. 1974. The biogeoclimatic subzones of Vancouver Island and the adjacent mainland based on climax vegetation (3d approximation). Department of Lands, Forests, and Water Resources, Victoria, B.C. 1:63 360; 106 x 77 cm; colored map.

Major, J. 1963. A climatic index to vascular plant activity. Ecology 44: 485-498.

McNaughton, K.G. and T.A. Black. 1973. A study of evapotranspiration from a Douglas-fir forest using the energy balance approach. Water Resources Research 9: 1579-1590.

McMinn, R.G. 1957. Water relations in the Douglas-fir region on Vancouver Island. University of British Columbia, Vancouver, B.C. 114 p. Dissertation.

Mueller-Dombois, D. 1959. The Douglas-fir forest associations on Vancouver Island in their initial stages of secondary succession. University of British Columbia, Vancouver, B.C. 570 p. Dissertation.

Mueller-Dombois, D. and H. Ellenberg. 1974. Aims and methods of vegetation ecology. John Wiley and Sons, New York. 547 p.

Orloci, L. 1961. Forest types of the Coastal Western Hemlock Zone. University of British Columbia, Vancouver, B.C. 206 p. Thesis.

________. 1965. The Coastal Western Hemlock Zone on the southwestern British Columbia mainland. Ecology of Western North America 1: 18-37.

Packee, E.C. 1974. The biogeoclimatic subzones of Vancouver Island and the adjacent mainland and islands. MacMillan Bloedel Forest Research Note. 9 p. (mimeo)

__________. 1976. An ecological approach toward yield optimization through species allocation. University of Minnesota, Minneapolis, MN. 740 p. Dissertation.

Resource Planning Division. 1976. Planning at the sub-unit level. Discussion paper no. 6. Ministry of Forests, Resource Planning Division, Victoria, B.C. 22 p.

Roehmer, H. 1972. Forest vegetation and environments on the Saanich Peninsula. University of Victoria, Victoria, B.C. 405 p. Dissertation.

_________. and V.J. Korelus. 1975. Biophysical mapping of Sooke Division, Pacific Logging Co. Ltd. Pacific Logging Co. Ltd., Victoria, B.C. 35 p. (mimeo)

Rose, M.F. and C. Grant. 1976. Remote station climate prediction model. Victoria, B.C.: Environment and Land Use Committee Secretariat (ELUC), Data Services Division. 8 p. (mimeo)

Schmidt, R.L. 1971. Mapping of forest productivity zones. Ministry of Forests, British Columbia Forest Service, Research Division, Victoria, B.C. 6 p. (mimeo)

Schofield, W.B. 1968a. A checklist of Hepaticae and Anthocerotae of British Columbia. Syesis 1: 157-162.

____________. 1968b. A selectively annotated checklist of British Columbia mosses. Syesis 1: 163-175.

Sochova, V.B. and T.I. Isachenko. 1973. Geobotanical mapping. Research and Microfilm Publications, New York. 97 p. Translated from the Russian.

Sukachev, V.N. and H. Dylis, eds. 1964a. Osnovy lesnoi biogeocoenologii. Nauka, Moscow. 514 p. In Russian.

____________________________. 1964b. Fundamentals of forest biogeocoenology. Oliver and Boyd Ltd., Edinburgh and London. 672 p. Translated from the Russian.

Tansley, A.G. 1935. The use and abuse of vegetational concepts and terms. Ecology 16: 284-307.

Taylor, R.L. and B. MacBryde. 1977. Vascular plants of British Columbia: a descriptive resource inventory. University of British Columbia, Vancouver, B.C. 754 p.

Thornthwaite, C.W. 1948. An approach toward a rational classification of climate. Geogr. Rev. 38: 55-94.

Trewartha, G.T. 1957. Elements of geography, physical, and cultural. McGraw-Hill Co., New York. 693 p.

____________. 1968. An introduction to climate. 4 edition. McGraw-Hill Book Co., New York. 408 p.

UNESCO. 1973. International classification and mapping of vegetation. United Nations Educational, Scientific, and Cultural Organization. Paris, France. 93 p.

Utzig, G. and D. MacDonald. 1977. Guide for tree species selection in the Nelson Forest District. (1st approximation). Ministry of Forests, British Columbia Forest Service, Victoria, B.C. 23 p.

Valentine, K.W.G. and L.M. Lavkulich. 1978. The soil orders of British Columbia. P. 67-95 in K.W.G. Valentine, P.N. Sprout, T.E. Baker, and L.M. Lavkulich, eds. The soil landscapes of British Columbia. Resource Analysis Branch, Ministry of Environment, Victoria, B.C.

Wade, L.K. 1965. Vegetation and history of the *Sphagnum* bogs of the Tofino area, Vancouver Island. University of British Columbia, Vancouver, B.C. 125 p. Thesis.

Wilson, R.G. and W.R. Rouse. 1972. Moisture and temperature limits of the equilibrium evapotranspiration model. J. of Appl. Meteorology 11(3): 436-442.

Wyeth, M.H. and K. Klinka. 1975. A refinement for species selection in planting. Ministry of Forests, British Columbia Forest Service, Victoria, B.C. Reforestation Notes 2(2): 6-7.

APPENDIX I

List of Climatic Observations Used in Discriminant Analysis

Climatic Data Stations	Elevations (m)	Characteristics of Prediction Points for Derived Climatic Data: Locations	Aspect (degrees from N)	Elevations (m)	Number of Observations
Agassiz, CDA	15				1
Alberni, Beaver Creek	91	Alberni, Beaver Creek	270	152 762 1067 1372	4
Alberni, Lupsi Cupsi	9				1
Alert Bay	52				1
Alouette Lake	117				1
Amphitrite Point	11	Amphitrite Point	270	457 762 1219	8
		Ucluelet	270	76 457 762 1067	
Ash River Generating Station	91	Ash River Generating Station	0	457 762 1067 1372	5
Bear Creek	351	Bear Creek	270	457 762 1067	3
Bonilla Island	16				1
Bowen Island, Arbutus Bay	23				1
Britannia Beach	49				1
Bull Harbour	14				1
Buntzen Lake	17				1
Burnaby Mountain Terminal	134				1

APPENDIX I (Continued)

Climatic Data Stations	Elevations (m)	Characteristics of Prediction Points for Derived Climatic Data: Locations	Aspect (degrees from N)	Elevations (m)	Number of Observations
Campbell River	79	Campbell River	90	76 310 457 762 1067	5
Chatham Point	79				1
Chilliwack	6	Sumas Canal	0	6	2
Clayoquot	23		270	457 762	3
Comox Airport	23	Comox Airport	90	76 457 762 1219	
		Courtenay O.A.S.	60	457 610 1372	7
Coquitlam	161				1
Cordova Bay	37	Cordova Bay	90	76	1
Cowichan Lake, Forestry	177	Cowichan Lake	90	152 762 1067 1372	4
Cultus Lake	46				1
Duncan	9	Duncan	135	76 762	
		Chemainus	90	229 762 1219	5
Estevan Point	6	Estevan Point	270	76 457 762 1067 1219	
		Tahsis	270	457 762 1067 1372	9
Haney, East	31				1
Haney, UBC RF	171				1
Holberg	569				1

APPENDIX I (Continued)

Climatic Data Stations	Elevations (m)	Characteristics of Prediction Points for Derived Climatic Data: Locations	Aspect (degrees from N)	Elevations (m)	Number of Observations
Hollyburn Ridge	951	Hollyburn Ridge	0	76	1
James Island	54	James Island	135	76	1
Ladner	1				1
Merry Island	58				1
Mission	56				1
Nanaimo	70	Nanaimo	90	76 152 762 1067 1372	
		Parksville	90	76 229 457 762 1067	10
Nanaimo, Departure Bay	8				1
New Westminster	119				1
New Westminster, BC Penitentiary	18				1
Ocean Falls	5	Bella Bella	0	12	
		Namu	0	49	3
Pachena Point	47	Bamfield	270	76 457 762 1067	
		Henderson Lake	270	457 762 1067	
		Kildonam	270	76 457 762 1067	
		Pachena Point	270	76 457 762 1067	15

APPENDIX I (Continued)

Climatic Data Stations	Elevations (m)	Characteristics of Prediction Points for Derived Climatic Data: Locations	Aspect (degrees from N)	Elevations (m)	Number of Observations
Pitt Meadows Lougheed	6				1
Port Alberni	59	Port Alberni	270	76 762 1067	3
Port Alice	15				1
Port Hardy Airport	23				1
Powell River	52	Powell River	270	76 229	2
Quatsino	2				1
River Jordan	3	River Jordan	0	76 762 1067	3
Saltspring Island	73	Pender Island	90	76 457	
		Saltspring Island	90	76	3
Saltspring Vesuvius	8	Saltspring Vesuvius	90	76 457	2
Shawnigan Lake	137	Sooke Lake	180	229 762	2
Spring Island	11	Kyuquot	0	11	2
Squamish	2				1
Vananda	12	Vananda	270	76	1
Vancouver International Airport	5				1

APPENDIX I (Continued)

Climatic Data Stations	Elevations (m)	Characteristics of Prediction Points for Derived Climatic Data			Number of Observations
		Locations	Aspect (degrees from N)	Elevations (m)	
Vancouver, UBC	87				1
Victoria, Gonzales Heights	69	Esquimalt	135	76	2
Victoria International Airport	20				1
White Rock	70				1
William Head	12	William Head	135	76	
		Sooke	180	229 762	3

APPENDIX II A

Means and Standard Deviations of Climatic Characteristics of the CDF, CWH, and MH Zones in Coastal Southwestern British Columbia

Climatic Characteristics *	Biogeoclimatic Zones		
	CDF	CWH	MH
Number of climatic data sets	32	79	27
Climate (Köppen/Trewartha)	Csb	Cfb	Dfc
Mean annual precipitation (mm)	1064 (243.4) **	2728 (1063.9)	3345 (1285.5)
Mean temperature of the coldest month (^{o}C)	2.5 (1.09)	1.0 (2.17)	-3.9 (1.88)
Index of continentality	12 (3.6)	9 (7.3)	10 (7.7)
Mean radiation during growing season (Ly)	44200 (2400)	39900 (2300)	39900 (1720)
Mean precipitation: April-September (mm)	248 (60.9)	671 (289.4)	760 (353.2)
Mean temperature of the warmest month (^{o}C)	16.9 (1.09)	14.2 (2.54)	9.6 (1.82)
Accumulated degree days over $5.6^{o}C$	1660 (182.4)	1165 (441.4)	348 (186.2)
Frost-free period (days over $0^{o}C$)	234 (35.3)	200 (36.2)	121 (20.7)
Mean precipitation of the driest month (mm)	26 (7.9)	69 (33.2)	102 (139.5)
Mean precipitation of the wettest month (mm)	176 (42.1)	413 (138.6)	527 (177.9)
Mean annual temperature (^{o}C)	9.5 (0.72)	7.5 (1.83)	2.7 (0.84)
Number of months with mean temperature over $10^{o}C$	5.4 (0.66)	4.0 (1.69)	0.8 (0.92)
Number of months with mean temperature below $0^{o}C$	0.0 (0.0)	0.5 (1.06)	4.7 (0.72)
Water surplus (mm)	678 (218.8)	2235 (968.3)	3028 (1271.4)
Water deficit (mm)	222 (69.2)	41 (54.6)	19 (36.9)

APPENDIX II A (Continued)

Number of months with water deficit	3.7 (0.74)	1.0 (1.13)	0.6 (0.84)
Maximum snow depth (cm)	5.9 (9.45)	59.5 (64.22)	294.2 (110.04)
Number of months with snow	0.6 (0.87)	2.9 (2.36)	7.7 (0.96)
Potential evapotranspiration (mm)	606 (36.8)	503 (46.6)	341 (55.7)
Actual evapotranspiration (mm)	386 (43.2)	463 (59.0)	322 (46.0)
Ratio actual to potential evapotranspiration (%)	64 (9.1)	92 (10.5)	95 (9.2)
Actual evapotranspiration: April-September (mm)	335 (49.3)	430 (49.1)	316 (43.6)
Actual evapotranspiration: February (mm)	4 (2.6)	2 (1.8)	0 (0.0)
Actual evapotranspiration: March (mm)	26 (3.2)	14 (11.1)	0 (0.0)
Actual evapotranspiration: April (mm)	56 (4.2)	40 (13.1)	0 (0.0)
Actual evapotranspiration: May (mm)	97 (5.1)	85 (5.7)	26 (19.1)
Actual evapotranspiration: June (mm)	76 (25.6)	97 (6.6)	76 (23.2)
Actual evapotranspiration: July (mm)	27 (12.4)	92 (27.6)	98 (18.3)
Actual evapotranspiration: August (mm)	34 (11.1)	67 (17.1)	69 (12.4)
Actual evapotranspiration: September (mm)	46 (7.0)	49 (5.1)	44 (4.7)
Actual evapotranspiration: October (mm)	18 (3.4)	15 (2.6)	7 (6.3)
Actual evapotranspiration: November (mm)	1 (1.3)	0 (0.0)	0 (0.0)

* Characteristics that are left justified in the columns were identified as significant variables in the discriminant analysis.

** Numbers in parenthesis are standard deviations.

APPENDIX II B

Means and Standard Deviations of Climatic Characteristics of the Maritime Forested and Parkland MH Subzones

Climatic Characteristics *	Biogeoclimatic Subzones: Maritime Forested MH	Biogeoclimatic Subzones: Maritime Parkland MH
Number of climatic data sets	18	9
Climate (Köppen/Trewartha)	milder Dfc	colder Dfc
Mean annual precipitation (mm)	3339 (1248.4) **	3358 (1434.6)
Mean temperature of the coldest month (°C)	-3.3 (1.60)	-5.0 (1.97)
Index of continentality	9 (7.8)	11 (7.8)
Mean radiation during growing season (Ly)	40100 (1780)	39500 (1590)
Mean precipitation April-September (mm)	762 (341.6)	755 (396.8)
Mean temperature of the warmest month (°C)	9.9 (1.91)	8.9 (1.53)
Accumulated degree days over 5.6°C	392 (196.3)	260 (132.9)
Frost-free period (days over 0°C)	127 (15.8)	109 (24.9)
Mean precipitation of the driest month (mm)	115 (168.0)	76 (45.8)
Mean precipitation of the wettest month (mm)	525 (171.8)	532 (200.4)
Mean annual temperature (°C)	3.2 (0.48)	1.7 (0.51)
Number of months with mean temperature over 10°C	1.1 (1.00)	0.3 (0.50)
Number of months with mean temperature below 0°C	4.5 (0.62)	5.1 (0.78)
Water surplus (mm)	2998 (1239.2)	3087 (1408.8)
Water deficit (mm)	27 (43.1)	3 (6.2)

APPENDIX II B (Continued)

Number of months with water deficit	0.8 (0.94)	0.2 (0.44)
Maximum snow depth (cm)	261 (77.6)	361 (138.1)
Number of months with snow	7.3 (0.89)	8.4 (0.53)
Potential evapotranspiration (mm)	376 (26.5)	272 (21.8)
Actual evapotranspiration: (mm)	348 (28.9)	269 (18.5)
Ratio actual to potential evapotranspiration (%)	93 (10.7)	99 (2.2)
Actual evapotranspiration: April-September (mm)	340 (30.9)	269 (18.5)
Actual evapotranspiration: February (mm)	0 (0.0)	0 (0.0)
Actual evapotranspiration: March (mm)	0 (0.0)	0 (0.0)
Actual evapotranspiration: April (mm)	0 (0.0)	0 (0.0)
Actual evapotranspiration: May (mm)	40 (2.6)	0 (0.0)
Actual evapotranspiration: June (mm)	92 (5.1)	44 (2.0)
Actual evapotranspiration: July (mm)	95 (20.9)	105 (9.1)
Actual evapotranspiration: August (mm)	67 (14.0)	74 (7.0)
Actual evapotranspiration: September (mm)	45 (4.3)	43 (5.5)
Actual evapotranspiration: October (mm)	11 (4.3)	0 (0.0)
Actual evapotranspiration: November (mm)	0 (0.0)	0 (0.0)

*, ** See footnotes in Appendix II A.

APPENDIX II C

Means and Standard Deviations of Climatic Characteristics of the West and East Vancouver Island Maritime Forested MH Variants

Climatic Characteristics *	Biogeoclimatic Variants	
	West Vancouver Island Maritime Forested MH	East Vancouver Island Maritime Forested MH
Number of climatic data sets	9	9
Mean annual precipitation (mm)	4358 (510.6) **	2320 (845.0)
Mean temperature of the coldest month (°C)	-2.0 (0.83)	-4.7 (0.81)
Index of continentality	2 (3.9)	16 (2.1)
Mean radiation during growing season (Ly)	39200 (1720)	41000 (1430)
Mean precipitation April-September (mm)	1063 (165.2)	460 (127.0)
Mean temperature of the warmest month (°C)	8.2 (0.89)	11.6 (0.60)
Accumulated degree days over 5.6°C	221 (91.1)	563 (88.6)
Frost-free period (days over 0°C)	134 (14.3)	120 (14.6)
Mean precipitation of the driest month (mm)	186 (220.2)	44 (8.9)
Mean precipitation of the wettest month (mm)	656 (67.6)	394 (139.9)
Mean annual temperature (°C)	3.0 (0.27)	3.3 (0.61)
Number of months with mean temperature over 10°C	0.2 (0.67)	1.9 (0.33)
Number of months with mean temperature below 0°C	4.0 (0.50)	5.0 (0.0)
Water surplus (mm)	4015 (528.3)	1981 (809.5)
Water deficit (mm)	0 (0.0)	54 (47.9)

APPENDIX II C (Continued)

Number of months with water deficit	0.0 (0.0)	1.6 (0.73)
Maximum snow depth (cm)	281 (42.7)	240 (100.1)
Number of months with snow	7.9 (0.33)	6.7 (0.87)
Potential evapotranspiration (mm)	354 (14.0)	398 (14.7)
Actual evapotranspiration (mm)	354 (14.0)	343 (38.8)
Ratio actual to potential evapotranspiration (%)	100 (0.0)	86 (11.9)
Actual evapotranspiration: April-September (mm)	346 (18.1)	334 (40.3)
Actual evapotranspiration: February (mm)	0 (0.0)	0 (0.0)
Actual evapotranspiration: March (mm)	0 (0.0)	0 (0.0)
Actual evapotranspiration: April (mm)	0 (0.0)	0 (0.0)
Actual evapotranspiration: May (mm)	38 (1.7)	41 (1.8)
Actual evapotranspiration: June (mm)	89 (4.9)	94 (3.9)
Actual evapotranspiration: July (mm)	99 (6.7)	91 (29.1)
Actual evapotranspiration: August (mm)	74 (4.9)	60 (16.7)
Actual evapotranspiration: September (mm)	42 (0.9)	48 (4.3)
Actual evapotranspiration: October (mm)	12 (2.4)	10 (5.5)
Actual evapotranspiration: November (mm)	0 (0.0)	0 (0.0)

*, ** See footnotes in Appendix II A.

APPENDIX II D

Means and Standard Deviations of Climatic Characteristics of the Drier and Wetter Maritime CDF Subzones

Climatic Characteristics *	Biogeoclimatic Subzones	
	Drier Maritime CDF	Wetter Maritime CDF
Number of climatic data sets	15	17
Climate (Köppen/Trewartha)	drier Csb	wetter Csb
Mean annual precipitation (mm)	907 (139.9) **	1202 (233.2)
Mean temperature of the coldest month (^{o}C)	3.2 (0.72)	2.0 (1.07)
Index of continentality	11 (3.2)	13 (3.9)
Mean radiation during growing season (Ly)	45700 (2510)	43000 (1430)
Mean precipitation April-September (mm)	214 (51.7)	279 (52.2)
Mean temperature of the warmest month (^{o}C)	17.1 (1.00)	16.6 (1.14)
Accumulated degree days over 5.6^{o}C	1753 (131.2)	1578 (185.1)
Frost-free period (days over 0°C)	253 (37.3)	217 (23.5)
Mean precipitation of the driest month (mm)	22 (8.0)	28 (6.9)
Mean precipitation of the wettest month (mm)	152 (23.6)	197 (43.5)
Mean annual temperature (^{o}C)	10.0 (0.40)	9.1 (0.72)
Number of months with mean temperature over 10°C	5.6 (0.51)	5.2 (0.73)
Number of months with mean temperature below 0°C	0.0 (0.0)	0.0 (0.0)
Water surplus (mm)	543 (111.4)	796 (223.2)
Water deficit (mm)	268 (70.4)	181 (34.9)

APPENDIX II D (Continued)

Number of months with water deficit	4.0 (0.65)	3.4 (0.71)
Maximum snow depth (cm)	1.0 (2.0)	10 (11.2)
Number of months with snow	0.2 (0.41)	1.0 (1.00)
Potential evapotranspiration (mm)	632 (35.6)	583 (18.7)
Actual evapotranspiration (mm)	364 (41.9)	405 (35.6)
Ratio actual to potential evapotranspiration (%)	58 (9.0)	69 (5.4)
Actual evapotranspiration: April-September (mm)	311 (48.6)	356 (40.5)
Actual evapotranspiration: February (mm)	5 (2.5)	4 (2.5)
Actual evapotranspiration: March (mm)	27 (3.7)	24 (1.9)
Actual evapotranspiration: April (mm)	59 (4.4)	54 (2.3)
Actual evapotranspiration: May (mm)	99 (4.7)	95 (4.7)
Actual evapotranspiration: June (mm)	60 (28.2)	90 (10.6)
Actual evapotranspiration: July (mm)	22 (8.0)	31 (14.1)
Actual evapotranspiration: August (mm)	29 (9.8)	39 (10.4)
Actual evapotranspiration: September (mm)	42 (7.2)	50 (4.6)
Actual evapotranspiration: October (mm)	20 (3.8)	17 (2.6)
Actual evapotranspiration: November (mm)	1 (1.2)	1 (1.5)

*, ** See footnotes in Appendix II A.

APPENDIX II E

Means and Standard Deviations of Climatic Characteristics of the Nanaimo & Georgia and Fraser Wetter Maritime CDF Variants

Climatic Characteristics *	Biogeoclimatic Variants	
	Nanaimo & Georgia Wetter Maritime CDF	Fraser Wetter Maritime CDF
Number of climatic data sets	11	6
Mean annual precipitation (mm)	1217 (257.7) **	1175 (199.7)
Mean temperature of the coldest month (^{o}C)	1.5 (0.82)	2.8 (1.01)
Index of continentality	14 (2.6)	10 (5.1)
Mean radiation during growing season (Ly)	42600 (1240)	43800 (1510)
Mean precipitation April-September (mm)	260 (44.3)	313 (51.2)
Mean temperature of the warmest month (^{o}C)	16.7 (1.14)	16.5 (1.24)
Accumulated degree days over 5.6^{o}C	1534 (196.8)	1660 (140.8)
Frost-free period (days over 0°C)	207 (18.3)	234 (23.0)
Mean precipitation of the driest month (mm)	27 (7.7)	31 (4.5)
Mean precipitation of the wettest month (mm)	207 (46.9)	179 (32.1)
Mean annual temperature (^{o}C)	8.8 (0.74)	9.6 (0.29)
Number of months with mean temperature over 10°C	4.8 (0.60)	5.8 (0.41)
Number of months with mean temperature below 0°C	0.0 (0.0)	0.0 (0.0)
Water surplus (mm)	826 (249.7)	742 (170.8)
Water deficit (mm)	192 (26.2)	162 (42.6)

APPENDIX II E (Continued)

Number of months with water deficit	3.8 (0.40)	2.7 (0.52)
Maximum snow depth (cm)	15 (10.8)	1 (1.6)
Number of months with snow	1.5 (0.93)	0.2 (0.41)
Potential evapotranspiration (mm)	581 (12.4)	588 (27.9)
Actual evapotranspiration (mm)	390 (28.8)	433 (30.6)
Ratio actual to potential evapotranspiration (%)	67 (4.4)	73 (5.2)
Actual evapotranspiration: April-September (mm)	344 (33.0)	379 (46.2)
Actual evapotranspiration: February (mm)	4 (3.2)	4 (0.5)
Actual evapotranspiration: March (mm)	24 (2.3)	24 (0.5)
Actual evapotranspiration: April (mm)	53 (1.3)	56 (1.9)
Actual evapotranspiration: May (mm)	94 (4.7)	98 (4.0)
Actual evapotranspiration: June (mm)	87 (10.5)	97 (7.9)
Actual evapotranspiration: July (mm)	27 (7.7)	40 (19.6)
Actual evapotranspiration: August (mm)	36 (10.9)	44 (7.6)
Actual evapotranspiration: September (mm)	48 (4.6)	54 (1.5)
Actual evapotranspiration: October (mm)	17 (3.3)	17 (0.0)
Actual evapotranspiration: November (mm)	1 (1.8)	0 (0.0)

*, ** See footnotes in Appendix II A.

APPENDIX II F

Means and Standard Deviations of Climatic Characteristics of the Drier and Wetter Maritime and the Wetter Maritime Northern CWH Subzones

Climatic Characteristics *	Biogeoclimatic Subzones: Drier Maritime CWH	Wetter Maritime CWH	Wetter Maritime Northern CWH
Number of climatic data sets	20	50	9
Climate (Köppen/Trewartha)	drier Cfb	wetter Cfb/c	wetter Cfb
Mean precipitation (mm)	1868 (268.7) **	3038 (1047.6)	2920 (1284.2)
Mean annual temperature of the coldest month (°C)	1.4 (0.91)	0.8 (2.60)	1.7 (1.31)
Index of continentality	15 (4.5)	6 (7.2)	7 (5.3)
Mean radiation during growing season (Ly)	40600 (2030)	40200 (1490)	36500 (3600)
Mean precipitation April-September (mm)	468 (68.2)	713 (261.8)	891 (466.4)
Mean temperature of the warmest month (°C)	17.3 (1.11)	12.9 (1.89)	14.5 (1.63)
Accumulated degree days over 5.6°C	1700 (204.9)	938 (335.9)	1241 (253.4)
Frost-free period (days over 0°C)	211 (19.3)	193 (41.5)	214 (21.6)
Mean precipitation of the driest month (mm)	46 (10.8)	74 (30.0)	96 (51.9)
Mean precipitation of the wettest month (mm)	291 (54.2)	461 (129.9)	416 (158.4)
Mean annual temperature (°C)	9.4 (0.76)	6.7 (1.65)	8.1 (0.91)
Number of months with mean temperature over 10°C	5.4 (0.51)	3.4 (1.75)	4.4 (0.73)
Number of months with mean temperature below 0°C	0.0 (0.0)	0.8 (1.23)	0.0 (0.0)
Water surplus (mm)	1383 (275.7)	2535 (896.7)	2460 (1266.0)
Water deficit (mm)	72 (47.5)	35 (56.7)	6 (11.1)

APPENDIX II F (Continued)

Number of months with water deficit	1.8 (0.72)	0.8 (1.18)	0.6 (0.88)
Maximum snow depth (cm)	19 (17.1)	80 (70.9)	38 (39.7)
Number of months with snow	1.4 (1.04)	3.7 (2.47)	2.1 (1.90)
Potential evapotranspiration (mm)	556 (19.9)	489 (36.1)	466 (51.2)
Actual evapotranspiration (mm)	483 (40.9)	455 (65.2)	460 (50.2)
Ratio actual to potential evapotranspiration (%)	87 (8.2)	93 (11.4)	99 (2.5)
Actual evapotranspiration: April-September (mm)	440 (39.4)	426 (52.5)	431 (50.3)
Actual evapotranspiration: February (mm)	3 (1.5)	2 (1.9)	1 (1.1)
Actual evapotranspiration: March (mm)	22 (1.2)	11 (11.9)	15 (8.9)
Actual evapotranspiration: April (mm)	50 (2.5)	35 (14.2)	45 (3.5)
Actual evapotranspiration: May (mm)	90 (5.6)	85 (4.5)	80 (6.4)
Actual evapotranspiration: June (mm)	103 (4.8)	97 (4.8)	89 (8.5)
Actual evapotranspiration: July (mm)	90 (31.4)	92 (27.9)	96 (17.6)
Actual evapotranspiration: August (mm)	59 (14.9)	69 (17.4)	76 (12.7)
Actual evapotranspiration: September (mm)	53 (4.5)	48 (4.0)	42 (5.5)
Actual evapotranspiration: October (mm)	16 (1.7)	15 (2.7)	13 (2.3)
Actual evapotranspiration: November (mm)	0 (0.0)	0 (0.0)	0 (0.0)

*, ** See footnotes in Appendix II A.

APPENDIX II G

Means and Standard Deviations of Climatic Characteristics of the East Vancouver Island and Pacific Ranges Drier Maritime CWH Variants

Climatic Characteristics *	Biogeoclimatic Variants	
	East Vancouver Island Drier Maritime CWH	Pacific Ranges Drier Maritime CWH
Number of climatic data sets	6	14
Mean annual precipitation (mm)	2060 (267.3) **	1785 (231.7)
Mean temperature of the coldest month (°C)	0.9 (1.12)	1.6 (0.76)
Index of continentality	15 (5.5)	14 (4.2)
Mean radiation during growing season (Ly)	42200 (2130)	39900 (1580)
Mean precipitation April-September (mm)	404 (38.5)	496 (59.3)
Mean temperature of the warmest month (°C)	16.8 (1.53)	17.5 (0.83)
Accumulated degree days over 5.6°C	1540 (213.9)	1769 (163.8)
Frost-free period (days over 0°C)	196 (16.4)	218 (17.2)
Mean precipitation of the driest month (mm)	36 (4.1)	50 (9.8)
Mean precipitation of the wettest month (mm)	347 (40.6)	267 (39.8)
Mean annual temperature (°C)	8.7 (0.61)	9.7 (0.62)
Number of months with mean temperature over 10°C	5.0 (0.0)	5.6 (0.50)
Number of months with mean temperature below 0°C	0.0 (0.0)	0.0 (0.0)
Water surplus (mm)	1626 (263.0)	1279 (213.7)
Water deficit (mm)	133 (17.3)	45 (26.5)

APPENDIX II G (Continued)

Number of months with water deficit	2.2 (0.41)	1.6 (0.76)
Maximum snow depth (cm)	35 (19.7)	12 (10.3)
Number of months with snow	2.2 (1.17)	1.0 (0.78)
Potential evapotranspiration (mm)	567 (19.4)	551 (18.8)
Actual evapotranspiration (mm)	432 (13.0)	505 (25.7)
Ratio actual to potential evapotranspiration (%)	76 (2.4)	92 (4.6)
Actual evapotranspiration: April-September (mm)	396 (16.3)	459 (30.0)
Actual evapotranspiration: February (mm)	1 (1.8)	3 (0.6)
Actual evapotranspiration: March (mm)	21 (0.9)	23 (0.7)
Actual evapotranspiration: April (mm)	50 (1.9)	50 (2.8)
Actual evapotranspiration: May (mm)	93 (4.8)	88 (5.3)
Actual evapotranspiration: June (mm)	106 (6.1)	102 (3.7)
Actual evapotranspiration: July (mm)	47 (8.1)	108 (15.3)
Actual evapotranspiration: August (mm)	45 (8.1)	64 (13.2)
Actual evapotranspiration: September (mm)	55 (7.6)	52 (1.5)
Actual evapotranspiration: October (mm)	14 (2.1)	16 (0.8)
Actual evapotranspiration: November (mm)	0 (0.0)	0 (0.0)

*, ** See footnotes in Appendix II A.

APPENDIX II H

Means and Standard Deviations of Climatic Characteristics of the Variants of the Wetter Maritime CWH Subzone

Climatic Characteristics *	Biogeoclimatic Variants: Estevan Submontane Wetter Maritime CWH	West Vancouver Island Submontane Wetter Maritime CWH	West Vancouver Island Montane Wetter Maritime CWH	East Vancouver Island Montane Wetter Maritime CWH	East Vancouver Island Submontane Wetter Maritime CWH
Number of climatic data sets	9	11	11	12	4
Mean annual precipitation (mm)	3016 (70.8) **	3819 (852.5)	3941 (411.9)	1892 (740.8)	1904 (728.6)
Mean temperature of the coldest month (°C)	4.6 (0.37)	2.1 (1.04)	0.1 (0.99)	-2.3 (0.96)	-1.2 (0.55)
Index of continentality	1 (0.2)	2 (4.2)	2 (3.7)	15 (3.1)	15 (2.4)
Mean radiation during growing season (Ly)	40300 (310)	39400 (940)	39600 (1570)	41400 (1740)	41500 (1190)
Mean precipitation April-September (mm)	774 (48.8)	876 (79.0)	967 (135.8)	376 (120.6)	373 (108.2)
Mean temperature of the warmest month (°C)	14.2 (0.72)	12.4 (1.54)	10.2 (0.98)	13.6 (0.63)	14.8 (0.80)
Accumulated degree days over 5.6°C	1390 (145.5)	920 (242.0)	502 (118.4)	879 (88.2)	1062 (129.2)
Frost-free period (days over 0°C)	263 (11.4)	206 (15.0)	170 (13.9)	150 (13.7)	173 (3.4)
Mean precipitation of the driest month (mm)	83 (6.1)	91 (16.9)	102 (20.3)	36 (10.2)	38 (5.3)
Mean precipitation of the wettest month (mm)	445 (20.5)	541 (19.3)	597 (55.1)	324 (121.6)	326 (113.6)
Mean annual temperature (°C)	9.2 (0.43)	7.1 (0.87)	5.0 (0.48)	5.4 (0.40)	6.4 (0.49)
Number of months with mean temperature over 10°C	5.1 (0.93)	3.8 (0.87)	0.6 (1.03)	3.7 (0.65)	4.0 (0.0)
Number of months with mean temperature below 0°C	0.0 (0.0)	0.1 (0.30)	0.4 (0.92)	2.7 (0.78)	1.2 (0.50)
Water surplus (mm)	2485 (73.4)	3072 (138.1)	3516 (421.3)	1512 (702.2)	1566 (671.9)
Water deficit (mm)	0 (0.0)	0 (0.0)	3 (8.7)	106 (52.0)	109 (35.8)

APPENDIX II H (Continued)

Number of months with water deficit	0.0 (0.0)	0.0 (0.0)	0.2 (0.60)	2.3 (0.65)	2.5 (0.58)
Maximum snow depth (cm)	0 (0.0)	38 (32.4)	139 (38.3)	130 (71.2)	99 (48.3)
Number of months with snow	0.0 (0.0)	2.5 (1.29)	6.0 (0.77)	5.5 (0.52)	5.5 (0.58)
Potential evapotranspiration (mm)	531 (10.7)	502 (8.1)	436 (15.0)	486 (23.0)	497 (27.7)
Actual evapotranspiration (mm)	531 (10.7)	502 (8.1)	433 (8.1)	381 (43.1)	388 (54.0)
Ratio actual to potential evapotranspiration (%)	100 (0.0)	100 (0.0)	99 (1.8)	79 (10.3)	78 (7.5)
Actual evapotranspiration: April-September (mm)	485 (8.6)	456 (10.0)	419 (9.2)	366 (46.0)	376 (51.7)
Actual evapotranspiration: February (mm)	4 (0.8)	4 (0.7)	0 (0.0)	0 (0.0)	0 (0.0)
Actual evapotranspiration: March (mm)	25 (1.2)	23 (1.2)	0 (0.0)	0 (0,0)	0 (0.0)
Actual evapotranspiration: April (mm)	52 (1.6)	48 (2.0)	23 (0.9)	24 (1.8)	18 (11.9)
Actual evapotranspiration: May (mm)	88 (1.8)	83 (2.1)	80 (3.3)	86 (4.9)	90 (2.4)
Actual evapotranspiration: June (mm)	102 (1.4)	96 (2.7)	93 (4.6)	96 (5.5)	99 (4.7)
Actual evapotranspiration: July (mm)	110 (1.5)	105 (2.0)	103 (5.1)	63 (33.7)	56 (27.5)
Actual evapotranspiration: August (mm)	83 (1.4)	80 (2.8)	75 (4.1)	45 (14.1)	55 (5.8)
Actual evapotranspiration: September (mm)	50 (1.4)	47 (1.1)	45 (1.4)	51 (4.2)	48 (8.7)
Actual evapotranspiration: October (mm)	17 (1.2)	15 (0.8)	14 (2.4)	15 (3.6)	11 (3.6)
Actual evapotranspiration: November (mm)	0 (0.0)	0 (0.0)	0 (0.0)	0 (0.0)	0 (0.0)

*, ** See footnotes in Appendix II A.

APPENDIX III

List of Plant Species

Scientific Name	Common Name
Abies amabilis (Dougl. ex Loudon) Forbes	Amabilis fir
Abies grandis (Dougl. ex D. Donn) Lindl.	Grand fir
Abies lasiocarpa (Hook.) Nutt.	Subalpine fir
Acer macrophyllum Pursh	Bigleaf maple
Achlys triphylla (J.E. Smith) DC.	American vanilla leaf
Aira caryophyllea L.	Silver hair grass
Alectoria sarmentosa (Ach.) Ach.	
Allotropa virgata T. & G. ex Gray	Candystick
Amelanchier alnifolia (Nutt.) Nutt.	Common Saskatoon
Apocynum androsaemifolium L.	Spreading dogbane
Arbutus menziesii Pursh	Pacific madrone
Arctostaphylos columbiana Piper in Piper & Beattie	Bristly manzanita
Athyrium distentifolium Taush ex Opiz	Alpine lady fern
Bazzania ambigua (Lindenb.) Trev.	
Bazzania denudata (Torr. ex Gott *et al.*) Trev.	
Bazzania tricrenata (Wahlenb.) Lindb.	
Blechnum spicant (L.) Roth	Deer fern
Boschniakia hookeri Walpers	Vancouver groundcone
Brodiaea coronaria (Salisb.) Engl.	Harvest brodiaea
Bromus carinatus H. & A.	California brome grass
Bromus vulgaris (Hook.) Shear	Columbia brome grass
Calamagrostis nutkaensis (Presl) Steud.	Pacific small reed grass
Caltha biflora DC.	Two-flowered white marsh-marigold
Camassia leichtlinii (Baker) Wats.	Great camas
Camassia quamash (Pursh) Greene	Common camas
Campanula scouleri Hook. ex DC.	Scouler's harebell
Carex nigricans C.A. Mey.	Black alpine sedge
Carex pensylvanica Lam.	Long-stoloned sedge

* Scientific names of vascular plants after Taylor and MacBryde (1977), of mosses and hepatics after Crum, Steere and Anderson (1973) and Schofield (1968a, 1968b), and of lichens after Hale and Culberson (1970).

APPENDIX III (Continued)

Carex rossii Boott in Hook.	Ross' sedge
Cassiope mertensiana (Bong.) G. Don	Mertens' mountain heather
Cassiope tetragona (L.) D. Donn	Four-angled cassiope
Ceanothus sanguineus Pursh	
Ceanothus velutinus Dougl. ex Hook.	
Cetraria subalpina Imsh.	
Chamaecyparis nootkatensis (D. Don) Spach	Yellow-cedar
Chimaphila menziesii (R. Br.) Spreng.	Menzies' pipsissewa
Cladothamnus pyroliflorus Bong.	Copperbush
Claytonia perfoliata Donn ex Willd.	Miner's lettuce
Clintonia uniflora (Schult.) Kunth	Blue-bead clintonia
Collinsia grandiflora Lindl.	Large-flowered blue-eyed Mary
Coptis asplenifolia Salisb.	Spleenwort-leaved goldthread
Cornus canadensis L.	Canadian bunchberry
Cornus nuttallii Aud. ex T. & G.	Western flowering dogwood
Danthonia californica Boland.	California oat grass
Delphinium menziesii DC.	Menzies' delphinium
Dicranum pallidisetum (Bail. ex Holz.) Irel.	
Disporum smithii (Hook.) Piper	Smith's fairybells
Dodecatheon hendersonii Gray	Henderson's shootingstar
Dryopteris assimilis Walker	Spiny shield fern
Erythronium oregonum Applegate	White fawn lily
Galium aparine L.	Common cleavers
Gaultheria ovatifolia Gray	Oregon-wintergreen
Gaultheria shallon Pursh	Salal
Hemitomes congestum Gray	Gnome-plant
Holodiscus discolor (Pursh) Maxim.	Oceanspray
Hookeria lucens (Hedw.) Sm.	

APPENDIX III (Continued)

Hylocomium splendens (Hedw.) B.S.G.	
Hypopitys monotropa Crantz	Fringed pinesap
Isopterygium elegans (Brid.) Lindb.	Synonym: *Plagiothecium elegans* (Brid.) (Sull.)
Isothecium stoloniferum (Brid.)	Synonym: *Isothecium spiculiferum* (Mitt.) Ren. & Card.
Juniperus scopulorum Sarg.	Rocky Mountain juniper
Listera cordata (L.) R. Br. in Ait.	Heart-leaved twayblade
Lobaria oregana (Tuck.) Müll. Arg.	
Loiseleuria procumbens (L.) Desv.	Alpine-azelea
Lomatium utriculatum (Nutt. ex T. & G.) Coult. & Rose	Fine-leaved lomatium
Lonicera hispidula (Lindl.) Dougl. ex T. & G.	Hairy honeysuckle
Lonicera utahensis Wats.	Utah honeysuckle
Luetkea pectinata (Pursh) Kuntze	Luetkea
Lycopodium alpinum L.	Alpine club-moss
Lycopodium annotinum L.	Stiff club-moss
Lycopodium sitchenses Rupr.	Alaska club-moss
Mahonia aquifolium (Pursh) Nutt.	Tall Oregon-grape
Mahonia nervosa (Pursh) Nutt.	Dull Oregon-grape
Menziesia ferruginea J.E. Smith	Rusty Pacific menziesia
Mimulus alsinoides Dougl. ex Benth.	Chickweed monkeyflower
Moehringia macrophylla (Hook.) Fenzl	Big-leaved sandwort
Mylia taylori (Hook.) S. Gray	
Navarretia squarrosa (Esch.) H. & A.	Skunkweed
Nemophila parviflora Dougl. ex Benth.	Small-flowered nemophila
Oemleria cerasiformis (H. & A.) Landon	Indian-plum
Oplopanax horridus (J.E. Smith) Mig.	Devil's-club
Orthocaulis floerkii (Web. & Mohr.) Buch.	

APPENDIX III (Continued)

Paxistima myrsinites (Pursh) Raf.	Oregon boxwood
Phyllodoce empetriformis (J.E. Smith) D. Don	Red mountain-heather
Picea sitchensis (Bong.) Carr.	Sitka spruce
Pilophoron clavatus Th. Fr.	
Pinus contorta Dougl. ex Loudon	Lodgepole pine
Pinus monticola Dougl. ex D. Don in Lamb.	Western white pine
Plagiothecium undulatum (Hedw.) B.S.G.	
Plectritis congesta (Lindl.) DC.	Rosy plectritis
Poa bulbosa L.	Bulbous blue grass
Polypodium scouleri Hook. & Grev.	Leathery polypody
Polystichum lonchitis (L.) Roth	Mountain holly-fern
Polystichum munitum (Kaulf.) Presl.	Western sword fern
Pseudoleskea baileyi Best & Grout ex Grout	
Pseudotsuga menziesii (Mirbel) Franco	Coast Douglas-fir
Pteridum aquilinum (L.) Kuhn in Decken	Bracken
Pyrola picta J.E. Smith in Rees	White-veined pyrola
Quercus garryana Dougl. ex Hook.	Garry oak
Rhizomnium glabrescens (Kindb.) Kop.	Synonym: *Mnium glabrescens* Kindb.
Rhizomnium nudum (Brit. & Williams) Kop.	Synonym: *Mnium nudum* Williams ex Britt. & Williams
Rhododendron albiflorum Hook.	White-flowered rhododendron
Rhytidiadelphus loreus (Hedw.) Warnst.	
Rhytidiadelphus triquetrus (Hedw.) Warnst.	
Rhytidiopsis robusta (Hedw.) Broth	
Ribes lobbii Gray	Gummy gooseberry
Rosa gymnocarpa Nutt. in T. & G.	Baldhip rose
Rubus pedatus J.E. Smith	Five-leaved creeping raspberry
Sanicula crassicaulis Poepp. ex DC.	Pacific sanicle
Scapania bolanderi Aust.	
Sisyrinchium douglasii Diet.	Douglas' blue-eyed-grass

APPENDIX III (Continued)

Smilacina stellata (L.) Desf.	Star-flowered false Solomon's-seal
Sorbus sitchensis Roemer	Sitka mountain-ash
Sphaerophorus globosus (Huds.) Vain.	
Stokesiella oregana (Sull.) Robins	Synonym: *Eurhynchium oreganum* (Sull.) Jaeg. & Sauerb
Streptopus roseus Michx.	Simple-stemmed twistedstalk
Streptopus streptopoides (Ledleb.) Frye & Rigg	Small twistedstalk
Symphoricarpos mollis Nutt. in T. & G.	Trailing snowberry
Thuja plicata Donn ex D. Donn in Lamb.	Western redcedar
Tiarella laciniata Hook.	Cut-leaved foamflower
Tiarella trifoliata L.	Trifoliate-leaved foamflower
Tiarella unifoliata Hook.	Unifoliate-leaved foamflower
Trachybryum megaptilum (Sull.) Schuf.	Synonym: *Homalothecium megaptilum* (Sull.) Robins
Tsuga heterophylla (Raf.) Sarg.	Western hemlock
Tsuga mertensiana (Bong.) Carr.	Mountain hemlock
Usnea longissima Ach.	
Usnea mollis Stirt.	
Vaccinium alaskaense Howell	Alaskan blueberry
Vaccinium deliciosum Piper	Cascade blueberry
Vaccinium membranaceum Dougl. ex Hook.	Black blueberry
Vaccinium ovalifolium J.E. Smith in Rees	Oval-leaved blueberry
Vaccinium ovatum Pursh	Evergreen huckleberry
Vaccinium parvifolium J.E. Smith in Rees	Red huckleberry
Vulpia microstachys (Nutt.) Munro ex Benth.	Small vulpia